Rock, Time, and Landforms

ROCK, TIME,

BY JEROME WYCKOFF

AND LANDFORMS

HARPER & ROW, PUBLISHERS, NEW YORK

PICTURE CREDITS

PHOTOGRAPHS:

American Airlines, 27; American Forest Products Industries, 241, 294, 296; American Geographical Society, 103 (Rich); Arabian-American Oil Co., 199, 213; Ausable Chasm Co., 74; Australian News and Information Bureau, 47, 114, 155; British Information Service, 165; British Travel Association, 23, 256, 257, 318; Bureau of Reclamation, Dept. of Interior, 187; Canadian Government Travel Bureau, 77, 123, 141, 234; Canadian Pacific Railway, 60; Chroussaki—Ruder and Finn, 122; Cold Regions Research and Engineering Laboratory, 56; Commonwealth of Puerto Rico, 293; Manfred d'Elia, 50; A. J. Eardley, 270; Ewing Galloway, 16; Finnish National Tourist Office, 240; Florida News Bureau, 191; Freeport Sulphur Co., 68; French Government Tourist Bureau, 290, 307; Geological Survey of Canada, 220; Hawaii Visitors Bureau, 168, 322, 340; H. M. Geological Survey (Crown Copyright Reserved), 15; Information Service of South Africa, 347; Italian State Tourist Office, 143; Japan Air Lines, 160; Japan Tourist Association, 20; Th. Josepsson, 337; Litton Industries—Aero Service Division, 4, 98, 196, 279; Meteor Crater Enterprises, Inc., 266; Montana Power Co., 263; Josef Muench, 32, 154, 185, 217, 259; National Park Service, 1, 6, 9, 36 (Anderson), 37, 40, 42, 55 (Boucher), 59, 72 (Anderson), 78, 79, 124, 125 (Grant), 152 *top* (Grant), 161 (Grant), 163, 164, 181, 183 (Grant), 186 (Grant), 195, 207 (Boucher), 208, 212, 214 (Boucher), 215 (Haugen), 247, 254, 255 (Palmer), 258 (Williams), 273, 283 (Kennicott), 285, 286 (Boucher), 287, 288 (Grant), 298 (Boucher), 303 (Boucher), 304, 308, 312 *top*, 314 (Anderson), 321 (Williams), 345; N. D. Newell, 330; New Hampshire Dept. of Resources and Economic Development, 28 (Dick Smith), 73 (Trask); New York State Dept. of Commerce, 95, 99; New Jersey Dept. of Conservation and Development, 311; New Zealand Government Travel Commissioner, 236; Northern Pacific Railway, 177 (Brown); Norwegian National Tourist Office, 245; Oregon State Highway Dept., 265; Royal Canadian Air Force, 239, 261; John S. Shelton, 86; Daniel Smiley, 111; Spanish Government Tourist Office, 2; Spence Air Photos, 312 *bottom*, 319; Bob and Ira Spring, 227; Standard Oil Co., 13, 64, 131, 189, 192; Swiss National Tourist Office, 142, 248; Texas Highway Dept., 344; Union Pacific Railroad, 87, 101, 139, 203; U.S. Air Force, 218; U.S. Coast and Geodetic Survey, 81, 156, 226, 302, 305, 315, 324; U.S. Forest Service, 10 (Prater), 100 (Higgins), 118, 147 (Monner), 148 (Prater), 152 *bottom* (Prater), 153 (Todd), 167 (Prater), 230 (Prater); U.S. Geological Survey, 44, 71 (Wentworth), 80 (Balsley), 116, 126, 145 (Balsley), 176 (Jones), 202 (Balsley), 209 (Piper), 267, 323 (Russell); U.S. Hydrographic Office, 341; U.S. Navy, 169, 229, 339; Harold Wanless, 108, 109, 278, 280; West Virginia Dept. of Commerce, 135 (Cruise).

Photographs not listed above were taken by the author

DIAGRAMS, CHARTS, AND MAPS:

The drawings were prepared by William Sayles.
Diagrams on the pages cited below were adapted by permission from the following works: 137—Douglas Johnson, *Stream Sculpture on the Atlantic Slope*, Copyright 1932, Columbia University Press, New York; 117, 237—A. K. Lobeck, *Geomorphology*, copyright 1939 by McGraw-Hill Book Company, New York; 299, 301, 309, 317, 333, 334—Francis P. Shepard, *Submarine Geology*, © 1963 by F. P. Shepard, Harper & Row, Publishers Incorporated, New York; 34, 112, 150, 244, 292—Arthur Strahler, *Physical Geography*, copyright 1951 by John Wiley & Sons, Inc., New York; 89—W. D. Thornbury, *Principles of Geomorphology*, copyright 1954 by John Wiley & Sons, Inc., New York. The chart on page 24 is reprinted by permission from *The Harper Encyclopedia of Science*, © 1963 by Harper & Row, Publishers, Incorporated, New York, and Sigma, Inc. The map on pages 62–63 is adopted from a more detailed map published by the U.S. Geological Survey.

ROCK, TIME, AND LANDFORMS. Copyright © 1966 by Jerome Wyckoff. Printed in the United States of America. All rights reserved. No part of this book may be used or reproduced in any manner whatsoever without written permission except in the case of brief quotations embodied in critical articles and reviews. For information address Harper & Row, Publishers, Incorporated, 49 East 33rd Street, New York, N.Y. 10016.

First Edition

Library of Congress Catalog Card Number: 66-10662
Designed by The Etheredges
Lithographed by The Murray Printing Company, Forge Village, Massachusetts

To Marjorie

Contents

Preface

1
Earth's Changing Face   1

2
The Meanings of Landscapes   7

3
Rocks, Air, and Water   31

4
Riverscapes   65

5
Up, Down, and Sidewise   105

6
A Gallery of Mountains   131

7
Volcanic Scenery   157

8
Plateaus and Plains   185

9
Realms of Wind   199

10
Ice on the Land   221

11
Lakes and Swamps   251

12
Limestone Country   275

13
Coasts and Shores   297

14
Continents, Islands, and Sea Bottoms   325

Further Reading   349

Glossary   351

Index   359

Preface

Traditionally, geology has been for geologists—the mining engineer, the geomorphologist, the petroleum prospector—and other specialists. Laymen have been little aware of the processes of erosion, volcanism, and crustal disturbance that have been shaping the Earth's surface for four thousand million years and are shaping it still. Mountain and plain, stream and canyon, lake and glacier are for most people only an unchanging backdrop for the drama of man and other living things.

This book about landscapes is for nonspecialists. It suggests new ways of looking at familiar scenes. It is for everyone who likes to look: the passenger in the jetliner and the spelunker in the cave, the hiker on the mountain and the stroller on the beach, the rockhound driving across the desert and the week-end sailor heading down a coast. Landforms are viewed here as drama in themselves, and their story, covering the enormous span of geologic time, is told in everyday terms.

To become more aware of landscapes is to gain a wider consciousness of the world; it is to share, in a sense, the physical existence of the planet itself. Certainly no time could be more appropriate than today for an expanded awareness of landforms. Human activities are altering scenery that has been millions of years in the making. Power dams and express highways, split-level developments and rebuilt beaches, swamp reclamation projects and irrigation canals, quarries and ski slopes—these are rapidly replacing natural features with the artificial.

Out of necessity man must and will continue to reshape his environment. Yet natural landforms are irreplaceable, and many are assets to be valued no less than Civil War battlefields and Egyptian temples. The conversion of wild and scenic lands for uses that are called progress, but often are unnecessary and destructive, is making all of us poorer—including the developers themselves. The more crowded our world becomes, the greater is our need to see things other than man and his works—things as they were before man came, and as they will be after he is gone.

Natural scenery is valued more as we come to know it better. People who are aware of how it evolves cannot but join in the struggle to protect it against needless change or obliteration. The future of rich, unspoiled portions of our natural heritage—the Grand Canyon and the Great Smokies, the Hudson Highlands and the Everglades, the Redwood stands and the Adirondacks, and all the others—depends upon a wider, deeper popular understanding of their intrinsic nature and their meanings to the human spirit. Perhaps this book will help toward such an understanding.

For technical advice generously given I am grateful to William D. Thornbury of Indiana University; to Maurice Schwartz, Nicolai Timofeef, and Lee Wilson of Columbia University; and to George Adams of the City College of New York. My thanks go also to William Sayles for his diagrams, to Alice Turner for her copyediting, and to Nancy Etheredge for her layouts. The long list of picture credits testifies to the helpfulness of the many sources that provided photographs, most notably the National Park Service, the U.S. Forest Service, the U.S. Geological Survey, and the U.S. Coast and Geodetic Survey.

J. W.

*In the Great Smokies, a raindrop glances on
the rock face, making a minute cascade.*

*On the bed of the Colorado River, a particle
of silt is lifted by the current, carried an
inch, and dropped again.*

*The dust devil in Utah blows another sand grain
down into Zion Canyon.*

*On the Maine shore, the ten-thousandth wave brings
the loosened block down from the granite cliff.*

*The root in the crevice, 11,000 feet high in the
Canadian Rockies, keeps pressing against the
rock walls.*

*Another molecule of pure water evaporates from
Great Salt Lake, leaving it saltier than before.*

*Seismographs record a slight earth tremor in the
vicinity of latitude 27 degrees north, longitude 87
degrees east. Mount Everest has risen another
0.0002 inch.*

*Two hundred feet below the floor of Mammoth Cave,
one more particle of calcium bicarbonate is
taken into solution by ground water.*

*By such events the land is shaped.
The ages are measured grain by grain.*

A world of geologic change: Mountain and lowland, lake and stream — all share in the unending process of destruction and renewal. Alaska's Mount McKinley and its neighbors, seen here across Wonder Lake, were built up by volcanic eruptions during recent geologic time and are now being steadily wasted by the elements.

I
Earth's Changing Face

To man, whose time is brief, the world and its landscapes seem everlasting and all but changeless. Rare is the observer who sees during his lifetime any thoroughgoing natural change in a mountain range or a river valley, a prairie or a seacoast. Land-forming processes are generally so gradual, so slow, that we are scarcely aware of them. The scholars of old took it for granted that Earth was created in one grand supernatural event, and that land and sea would last forever—or at least until the Day of Judgment. The early geologists, being unaware of the planet's vast age, could scarcely credit grain-by-grain processes such as weathering and erosion with the shaping of major landforms. Most laymen of our time, though taught in school that Earth was made by natural forces, are still but faintly aware of landscapes as features that do change radically through geologic time and are, in fact, being shaped today by the same forces that have shaped them in the past. We still speak of terra firma and the everlasting hills.

Earth's crust is in truth a scene of continuous, usually agonizingly

Sea against land: As at many locations elsewhere in the world, the sunken coast at Tossa de Mar, in Gerona, Spain, is being cut away by waves and currents.

slow, yet sometimes rapid, transformations. The rock that makes up the crust is under incessant, disintegrating chemical attack by the atmosphere and the waters of land and sea. It is riven by freezing moisture, by the expansive growth of plants, by strains due to changes in its temperature and volume. It is dissolved away by rainwash and by percolating water underground; it is scoured by sediment-laden streams, broken by the impact of waves, gouged and ground by glaciers. And it is disturbed, meanwhile, by powerful forces originating at depth. Currents of semiplastic material in the planet's hot interior, rising and then turning down again like air circulating about a radiator, distort and fracture the crust. As lofty highlands are worn down and as water and wind sweep the weighty rock waste to lower levels elsewhere, the great crustal blocks rise and sink uneasily, seeking a new balance. Under irresistible forces of tension and compression, the rock strata warp, fold, and break, and through the fissures molten stuff rises from the depths and bursts out at the surface in displays of volcanism.

All the terrains of the world—desert and jungle, plain and badland, mountain and valley—are at this moment being wasted by the elements. Streams are cutting valleys into the bedrock, and the sides of these valleys are being worn back and lowered by slopewash and weathering. The world's natural lakes, mostly relics of the last ice age, are being drained as outlet streams slice down through the sides of their basins. Along seacoasts, waves and currents are undercutting the cliffs and reforming the bars and beaches. Wind and water are cutting away the desert highlands and carrying the refuse down onto the lowlands. In humid highlands where the bedrock is of limestone, the solution process is making tunnels and caverns, gradually reducing the crust. Glaciers meanwhile are whittling the high peaks and gouging out the intermontane valleys, while ice sheets grind down the polar lands.

Erosion processes are usually slow-working. They may take a hundred thousand years to cut a deep ravine into a granite mountain, two million to cut and widen a Grand Canyon, and twenty-five million to level a mountain range like the Appalachians. But geologic time is time enough.

While some features are being worn down, others are being built up. Streams and sheetwash, assisted by wind, carry the debris of erosion pebble by pebble, grain by grain, to lower and lower levels. Some of the sediments pile up in deltas and deserts, or make valley fillings, or are spread broadly to form alluvial plains. Eventually they may consolidate to become rock again, and they may be raised up to form new highlands, so that the cycle of erosion can be repeated. Other sediments keep moving downward until they are dumped into the sea. But these sediments, too, may form rock again and, in some future eon, may be heaved up to become dry land.

The crust everywhere is unresting. The rate of movement is slow, ranging from virtual "stillstand" to shifts of inches or feet per century. But in geologic time just an inch per century is rapid enough to lift a Himalaya

An erosional remnant on the Great Plains: Scotts Bluff, Nebraska, is the last of a series of rock strata that once covered this region.

range miles above sea level, or to drop the bottom of a basin, such as that which holds Siberia's Lake Baikal, thousands of feet below it. Great active rifts recently discovered in the ocean bottoms hint that the continents themselves may be drifting.

Along lines of weakness in the crust nearly five hundred volcanoes in our day have been intermittently erupting. Volcanism in the Azores, the vicinity of Iceland, and around the margins of the Pacific is creating new lands before men's eyes, and on some of the older land masses new eruptions are continuing a building job begun long ago.

Even the sea bottoms are not exempt from change. Once believed to be quiet, undisturbed plains beneath the miles of dark water, they are now known to be evolving much as the land is evolving. The continental shelves and slopes, and the deep ocean floors too, are being planed, channeled, and canyoned by water in motion. Lava is piling up around vents in these bottoms. Escarpments, rifts, domes, even major mountain ranges, complete with valley systems, are familiar to the ocean floors.

The natural scenes around us were not created for eternity but are the images of a moment in geologic time. Though ephemeral they hold a weighty meaning. Where does the past appear more awesomely than in the ruins of the land? Where is the future more plainly in the making? Earth's crust is the wreck of worlds long vanished and the foundation of worlds to come. The massive hill and the winding stream, the curving dune and the turreted canyon, the sleeping volcano and even the multitudinous sand grains on the shore—all yield clues to four billion years of history and, it may be, to an equal span of future change.

II

The Meanings of Landscapes

For early man, landscapes were part of the fabric of life. Mountains were the abodes of superior beings, and volcanoes expressed their bad temper. Rivers were personified as life-givers, though a friendly stream could—and often did—swell up with anger and drown a village. The man- or animal-like profiles of rocky cliffs stood guard over the community. Out of the grand panorama of landforms arose the Great Spirit itself.

Civilized man long ago ceased to personify landforms, except in poetry. Yet the recognition of these features for what they are, the understanding of how they came to be, has come but recently. For thousands of years man lived on Earth's crust—hunting on it, tilling it, quarrying it, painting it, writing poems about it, philosophizing about it—with only the slightest awareness of what natural processes were doing to it day by day. Landforms were too familiar, perhaps, to provoke many questions; and when questions were asked, scholars were ready with answers that originated more from imagination than from observation and analysis.

Eroded pinnacles in Bryce Canyon, Utah: These features have been carved mostly by the action of rainwater working down through vertical joints in bedrock.

The state of knowledge concerning Earth's crust until recent times is fairly represented by the teachings of Aristotle. This renowned philosopher of classical Greece, thrusting aside the old myths, tried to reason about the world in terms of natural phenomena. Assuming the sun to be the most powerful force in the universe, he decided that solar heat had made the land by congealing great volumes of water. He reasoned further that mountains have been drawn up by the power of sun and stars, and also thrust up by subterranean winds and fires—which, incidentally, he credited with causing volcanic eruptions as well. The waters of the world were supposed to result from the condensation of vapors rising from Earth's interior and, to a lesser degree, from the condensation of air to form rain. Vapors bursting out through vents in the mountaintops condensed into water, which then drained down through ready-made valleys to the sea.

Some of the philosopher's ideas are suggestive of modern scientific fact. Most of the sedimentary rock strata that cover so much of the planet did indeed form from sediments laid down in water, though not from the water itself. Aristotle was much closer to the fact than later Greeks and medieval scholars who taught that rock, like plants and animals, grows from seeds. His further notion that the oceans are condensates of vapors erupted from Earth's interior brings to mind the similar modern theory of ocean formation during the planet's very early days. Yet it must be said that the classical Greeks and medieval thinkers, however accomplished otherwise, contributed little to our understanding of landforms. The basic processes—erosion, volcanism, and diastrophism, or crustal distortion—were witnessed and discussed, but were not in any important sense understood. Understanding could come, as in other areas of scientific knowledge, only as men came to appreciate the importance of meticulous observation and disciplined analysis in the tracing of natural causes.

Erosion: Not from Noah's Flood

Among the ancients, the first to sense the importance of erosion appears to have been the Greek historian and traveler Herodotus, who lived about 500 B.C. With truly startling understanding he speculated that the Nile Delta had been built up in a former gulf of the Mediterranean, from sediments carried by the Nile from highlands to the south, over a period of twenty thousand years or so. About five centuries later the Greek geographer Strabo likewise wrote of the power of rivers to carry sediments and deposit them to form vast flood plains and deltas. But such perceptions were infrequent and isolated, and were lost during the Middle Ages. Seventeen hundred years after Strabo, some natural philosophers in England, France, and Germany were explaining deltas, flood plains, and other works of erosion and deposition as the results of Noah's flood. This majestic pun-

Carved out by streams: Steamboat Rock, in Dinosaur National Monument, is typical of blocks carved out of high, arid plateaus by swift streams. The Green and Yampa rivers, which did the cutting, meet behind this monolith.

ishment imposed by God upon his corrupted children appeared to account not only for extensive sedimentary deposits but also for the presence of marine fossils in highlands far above the sea.

One of the first observers to sense the real magnitude of Earth's age and to realize that erosion has been a prime landscape-maker was the French naturalist Jean Guettard (1715-86). It was apparent to him that sedimentary strata, which lay hundreds and even thousands of feet thick over the land, had formed from erosion debris and organic remains. Only over tens of thousands of years, Guettard knew, could such deposits have been built up by ordinary erosion processes. Observing shore erosion along the French coasts, and noting that many sedimentary rocks—including formations far inland—contain marine fossils, he concluded that in the past large land areas must have been cut back and overwhelmed by ocean waves and currents.

First to recognize air and running water as the dominant sculptors of the crust was the Scottish farmer-geologist James Hutton (1726-97). This painstaking investigator and analytical genius saw the full significance of disintegrating rocks, of gullies in mountainsides and farmers' fields, of

ROCK, TIME, AND LANDFORMS · 10

sediment-laden streams winding through valleys toward the sea, and of the sedimentary rocks that covered most of the British Isles and Europe. Today's landscapes, Hutton realized, are the ruins of others that date back to an unimaginably remote past. In the slow, grand processes of change he saw, as he is reported to have said, "no trace of a beginning and no prospect of an end."

Hutton's appraisal of weathering and erosion, and his belief that the land was shaped in the past by much the same agents that are shaping it today, opened an age of rapid progress in geology. Among the great names of the time was that of Leopold von Buch (1774-1853), a widely traveled German who was the first to comprehend the vast role of volcanism in Earth's history. An Englishman, Charles Lyell (1797-1875), traveled widely through Europe and North America, studying particularly the evidences of land degradation by erosion and of land uplift by forces still completely mysterious. In America there was the soldier-explorer John Wesley Powell (1834-1902), whose explorations of the Colorado River region revealed the absolute dominance of streams in the degradation process. And there was Powell's contemporary, Louis Agassiz (1807-73), the imaginative Swiss,

Erosion by glaciers: Millions of square miles of Earth's crust have been sculptured by glaciers. This is North Sawyer Glacier in Alaska.

who recognized signs of former glaciation in the moraines, the scratched rock surfaces, and the widely traveled boulders scattered over the northern portions of Europe and North America, and who announced rightly that within recent geologic time most landscapes of temperate and arctic regions had been overwhelmed by grinding ice. To such pioneers is owed our modern understanding of the natural forces that make landforms.

Volcanism: The Fire in the Earth

Among all natural phenomena, volcanic eruptions probably have inspired the greatest wonder, if not always the greatest fright. The looming mountain with its plume of steam, the glowing rivers of lava on its slopes, the scorched and ash-mantled landscapes, the smoking fissures—all present a spectacle that is otherworldly indeed. The sight of a volcano in action is not soon forgotten.

Volcanism gave the mythmakers and philosophers of old much to explain. Among primitive peoples, legends tell of the time the mountain blew up, and the spirit rushed forth in fire to lay the world waste. In classical mythology Vulcan, the Roman god of fire who was also the blacksmith of the gods, was identified with an active, very noisy volcano on the island of Vulcano, north of Sicily. It was Vulcan after whom not only this island but all the processes by which volcanoes come into being were named. On the other side of the world, in Japan, beautiful Fujiyama was the holy mountain, inhabited by deities who might at any moment burst forth in anger. The spirit of the Hawaiian volcanoes was the fire goddess Pelé, who resided in the blazing caverns beneath Kilauea and who to this day appears to Hawaiians before an eruption.

Twenty-three centuries ago Aristotle attributed volcanoes to fires started by pent-up winds in subterranean deposits of sulfur and coal. But in the Middle Ages the alchemists and philosophers rallied around a grander concept: that volcanoes originate from a great fire at Earth's center, kept going by concentrated light from the stars. This central fire, exerting intolerable pressure on the planet's crust, accounted for volcanic eruptions and earthquakes—and, incidentally, provided a sobering reminder for sinners.

Among the ancients were, however, some careful observers. About 35 B.C. the Roman architect Vitruvius, reporting on the hot springs, steaming fissures, and "scorched" rocks around Mount Vesuvius, suggested that this peaceful mountain had once been the scene of terrible conflagrations. A half century later Strabo, the Greek geographer who had observed the constructive work of rivers, positively identified Vesuvius as a volcano, although local records made no mention of any past eruptions. It was in A.D. 79 that the side of Vesuvius blew out, engulfing Pompeii and Herculaneum in clouds of poisonous gas and dust, hot pumice, and steaming mud.

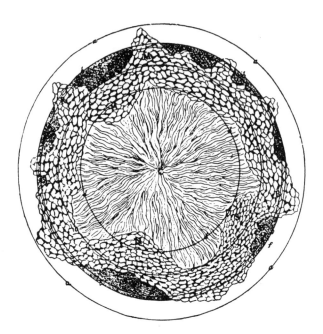

The great Central Fire: This diagram from the Italian Abbé Moro's *De crostacei* (1740) was intended to show how pressure from a great fire in Earth's interior has pushed up portions of the crust to form mountains.

Pliny the Elder, the great Roman natural philosopher, hurried to the scene in time to become an early martyr to scientific observation. Venturing ashore from his ship to watch more closely, he stayed too long.

In the eighteenth century, the age of reason, the great central fire still burned, but now it was a chemical phenomenon—the result of chance mixing of inflammable materials in the crust: coal, sulphur, niter, metallic sodium. Observers ranging over Europe noted many signs of the fire. Old cratered mountains and "scorched" domes or knoblike hills were studied and identified as, apparently, ruins of old volcanoes. Such sights prompted a disturbing question: Were these the evidence of a fiery creation of the world?

About 1800 Leopold von Buch, the first great volcanologist, decided that volcanoes result not from any central fire but rather from the upsurge of molten rock from separate reservoirs deep in Earth. The molten stuff forces the crust upward, he said, to form domes, some of which burst at the top and become craters, while others after cooling remain as solid rock masses. Later, Charles Lyell of England recognized cratered volcanic cones as accumulations of erupted materials around a vent, and he rightly identified Von Buch's uncratered domes as cores of extinct cones that have nearly eroded away.

Evidences of past volcanism were found widely distributed in Europe, North America, and Africa. Innumerable towerlike hills in certain regions, as in central France and Scotland, were identified as old volcanic cores. In the same regions, some ridges and escarpments were found to consist of resistant lava rock. Entire plains and plateaus were recognized as the surfaces of cooled lava flows. In some localities the products of recent volcanism were being rapidly removed by erosion, and in others erosion was exhuming volcanic remains of a remote past, long buried under sediment turned to rock. If all these relics did not prove, as some people feared, that Earth had begun in fire, they showed conclusively that igneous activity (as geologists were now calling it) had in the past played a role widely over the planet.

Diastrophism: No Terra Firma

The natural philosophers of the Mediterranean region have often been reminded of the instability of Mother Earth. During the human era this area has seen not only spectacular volcanic outbursts, such as those of Vesuvius, but severe earthquakes also, as the ruins of Greece and other ancient Mediterranean sites testify. It is little wonder that the old scholars talked of fires and winds raging in the realm of Pluto.

Yet as late as the eighteenth century learned men were underestimating the potency of diastrophism—that is, the distortion and fracturing of the crust by natural forces—as a landscape-maker, just as they were underestimating the powers of erosion. In the religious climate of the time, Noah's flood explained for many the great folds and breaks that are seen in some rock outcrops, particularly in mountain regions, just as it accounted for deltas and flood plains and the marine fossils in highland rocks. As long

Once the scene of volcanism: At Le Puy, in the Auvergne area of south-central France, the cores of extinct volcanoes were favored sites for churches.

Where great underground forces worked: Rock folds such as this syncline in a road cut near Wurtsboro, New York, suggested to early natural philosophers that Earth had suffered great cataclysms in the past.

as the world was considered to be only a few thousands of years old, nothing but the most tremendous cataclysms seemed to account for rugged scenery.

As with erosion, the results of diastrophism could be recognized only within the proper perspective. Within geologic time as James Hutton conceived it, an uplift rate of mere inches per century was enough to make lofty mountains. A sea bottom uplifted would explain more plausibly than the Deluge the marine fossils in rocks of the dry land. Earthquakes seemed obviously associated with the fracturing and dislocation of strata in rugged terrains, but cataclysms were unnecessary as causes. Not cataclysms, but rather slow, gentle uplifts seemed to explain the coverage of wide areas of the British Isles and Europe with only slightly disturbed sedimentary strata containing those telltale marine fossils.

The peripatetic Lyell and his successors found nearly everywhere some evidence of the diastrophic past. Along Baltic seacoasts, for example, they identified old beaches fifty to a hundred feet above sea level. Coastal areas of southeastern England also looked suspiciously like raised beaches. In the Mediterranean, certain coastal towns were known to have sunk beneath the waves during historic times. All in all, such evidence could mean merely that the sea had subsided or risen, not the land; but there was other evidence convincing beyond doubt. In highlands such as the Alps, the sequences of the rock strata showed severe displacements—sure signs of the movement of crustal blocks and slabs. Certain basins in the land, such as that of the Dead Sea, could be identified as terrains of subsidence. Even in regions where the surface sedimentary strata were nearly horizontal, outcrops in cliffs revealed that these strata were underlain, in some instances, by much-fractured and much-distorted formations. Clearly, the long history of the crust had been one of stress and strain, of slow but continual and extensive movements.

Toward the close of the nineteenth century it became clear that diastrophism not only produces certain landforms directly but also fundamen-

An uplifted Scottish shore: Early British geologists recognized raised wave-cut platforms as evidences of diastrophism. This one, 25 feet above the present level of wave work, is on the island of Kerrera, below Gylen Castle, Argyllshire.

A sunken terrain: The Valley of the Jordan River, east of the Mediterranean Sea, was recognized by early geologists as a great trough formed by faulting. The Sea of Galilee (shown here, with Tiberias in foreground) and the Dead Sea lie in this trough, far below sea level.

tally governs the rate of erosion. John Wesley Powell's concept of base level—the lowest level to which a land can be reduced by running water—was the key to this understanding. Reflecting upon the canyon-cutting by the swift streams on the recently uplifted Colorado Plateau, Powell realized that uplift is a prerequisite to erosion. Streams can reduce a terrain only if they have enough gradient, or slope, to cut down through the rock and carry the rock waste to lower levels. As a land is reduced, its streams lose cutting and carrying power, and the rate of reduction becomes slower and slower. As the gradient of the streams approaches zero, so does the rate of erosion. Erosion can resume only if the land is uplifted and the streams regain their vigor of flow.

Powell's concept, which is still basic today, had important meanings for the whole science of geology. And it raised at least one tremendous question. If Earth is so very old, and if weathering and erosion are as potent as the evidence indicates, why were not all the continents long ago wasted down to the ultimate base level—the level of the sea? The answer was obvious and certain: in the course of Earth's history the great land masses have been cut low again and again, but diastrophism has always restored them.

The Geologic Vista

Something less than two centuries of geologic science, along with accumulating knowledge in astronomy, has given us today a vista of Earth's long past. The planet is now believed to be about 4700 million years old. Probably it originated, along with the sun and other members of the solar system, from a tremendous swirling cloud of gas and dust in space. According to the now widely accepted "big-bang" theory, this cloud was produced by the explosion of a capsule of highly concentrated primordial matter.

Eddies in the original cloud gradually contracted as a result of gravitational pull between the drifting particles and also, perhaps, because of the pressure of radiation from surrounding stars and other bodies. The greatest of the eddies became the sun. This contained so much matter, with such great gravitational effect, that the contraction caused intense heating. The heat kept the sun in gaseous form, and it started the chain of thermonuclear reactions—the conversion of hydrogen into helium—that has to this day produced the sun's radiant energy.

The smaller aggregations of eddying particles in the primordial cloud became the planets and lesser bodies of the solar system. In these less heat was generated by contraction, and spheres that were solid or semisolid resulted—among them Earth. Just when Earth's solid crust formed is not known, but probably the interior was melted early by heat from contraction and radioactivity. Gases bursting through the surface produced the atmosphere, part of which condensed to form the oceans. The continents were built up perhaps by volcanic eruptions of rock-forming materials from ocean bottoms—eruptions probably like those that have formed the volcanic Hawaiian and Philippine islands, among others, in recent time.

With the emergence of the continents, the evolution of lesser landforms began. Fresh lava rock, on rising above the waves, was attacked by atmospheric gases and water, and eroded by streams and winds bearing sediments which acted as abrasives. Erosion debris accumulating in hollows on the land and in troughs between island chains became consolidated to form rock. This rock, typically lighter than the basalt rock that forms the bulk of the crust, tended to rise and to form new land, because of the balancing process in the crust—the process which, by the action of gravity, keeps the planet's mass about equally distributed around its center. As more and more lava was erupted, and as more and more sedimentary rock formed on and around the continents, these land masses kept growing.

This growth was neither continuous nor uniform. Volcanism waxed and waned and waxed again, occurring along lines of weakness in the crust. Diastrophic activity, which is usually associated with volcanism, also waxed and waned. As lands subsided and rose again, the sea invaded them and then withdrew, over and over. Erosion was most rapid when the lands were highest, and slowest when they were near base level. These patterns of change, established probably during the first billion years of the planet's existence, are essentially those that still prevail.

The Restless Crust

Our understanding of crustal change comes in part from what has been learned, indirectly, about Earth's interior. The existence of a magnetic field around the planet suggests that beneath the crust there are masses of liquid or semisolid material which rotate at speeds different from that of the relatively rigid crust. The refraction, or change in direction and speed, of seismic (earthquake) waves as they pass through the planet, before being recorded at seismographic stations, indicates that they have passed through "shells" of different densities. Measurements of gravity at different locations around the world reveal variations in the density of the crust. Analyses of lavas, some of which are believed to have been erupted from depths of fifty miles or more, yield information about the chemical nature of the interior.

From such studies a cross section of Earth has been visualized. At the center is the so-called inner core, a sphere apparently of nickel-iron, about 1600 miles in diameter. Encompassing it is the outer core, a shell of nickel-iron in a liquid or semisolid state, about 1300 miles thick. Next is the mantle, an 1800-mile-thick shell of somewhat plastic material which in chemical composition is believed to resemble the dark, heavy rock called peridotite. Finally, containing Earth's less-than-solid insides much as an eggshell contains an egg, is the relatively thin crust, a moderately rigid but much-fractured layer of bedrock averaging about twenty miles in thickness.

The materials of our planet are arranged approximately according to their respective densities, the heaviest being concentrated toward the center and the lightest in the crust. This distribution is not, however, completely stable. In the mantle particularly, heat from the decay of radioactive sub-

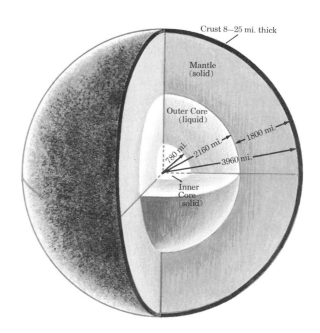

Earth's structure: The nearly rigid crust—the matrix of all landforms—is so thin as to be difficult to show in scale on a small drawing.

stances and perhaps from other sources, too, causes adjacent material to expand and thus sets up convection currents. Such currents though extremely sluggish are probably somewhat responsible for crustal rumplings that make mountains and plateaus, troughs and basins. Some crustal disturbance probably results also from variations between the rotational speed of the crust and that of the mantle.

Outside agencies, too, deform Earth's thin skin. Continuous changes in the relative positions of Earth, Moon, and Sun cause corresponding changes in gravitational pulls upon the planet. The crust, like the sea, has its changing tides. But even more important, probably, are the unbalancing effects of erosion. As highlands are destroyed, the rock debris drifts down into troughs and basins, there to accumulate to depths of hundreds or thousands of feet—even miles. Such transfers of weight on the crust cannot but produce profound consequences. Erosion is undoubtedly one of the most powerful causes in the drama of diastrophism.

If crustal change is constant and pervading, one may ask why the sizes and locations of the largest landforms—that is, the continents—appear to be so well established. The answer, as already suggested, lies in the fact that the crust consists mostly of two broad classes of rocks: the basalt-related varieties, which are similar chemically to the material of the mantle and which form a miles-thick shell around it; and the granite-related and sedimentary rocks, which generally cover the basaltic shell. The nonbasalts are mostly lighter in weight than the basalts and therefore, when accumulated in very large masses, tend to stand higher in the crust. The continents can be regarded as massive blocks of nonbasalts "floating" in basalt. The highest continental elevations are those beneath which the rocks are lightest, and the deepest ocean basins lie where the rock is heaviest. By the same token the crust is thickest beneath the continents and thinnest beneath the seas. This relationship between continents and ocean bottoms dates far back into the past. In recent geologic time there has been no great change in the locations of large land masses, although volcanic activity has continued to enlarge them, particularly along their margins.

The plastic nature of the mantle, the fractured state of the crust, and disturbances of both mantle and crust account for volcanism. As crustal blocks move up, down, and against one another, mantle material under the tremendous weight of the blocks squeezes upward through fissures. Melting the adjacent rock as it rises, this molten matter, known as magma, may cool and solidify within the crust to form so-called intrusive igneous rock. But on occasion it bursts through to the outer world, and then we have the spectacular phenomena of volcanic activity—exploding mountains, geysers, hot springs, and lava flows that may spread over thousands of square miles. Meanwhile, masses of intrusive igneous rock, when exposed by weathering and erosion, become responsible for the formation of such features as ridges, buttes, cliffs, and gorges.

The role of volcanism: Eruptions of matter from Earth's interior have built up much of the crust. The locality here is Akan National Park, Hokkaido, Japan.

Notwithstanding what has been said about the fragility and continuing disruption of the crust, this structure must be recognized as—by human standards—strong and solid. Because of its stupendous weight, estimated at nearly three tons per square inch at a depth of only a mile, sizable natural cavities cannot exist for long at depths of more than a few thousand feet. At depths of more than a mile or two any cracks produced by crustal movements will soon be either closed by pressure or filled by injections of magma.

The crust has not been distorted and disrupted so as to produce any really great extremes of relief, relatively speaking. With due respect for the glory of the Himalayas, the Alps, and other sky-splitting scenery, the most imposing landforms when seen in perspective are minuscule. The maximum relief achieved by any mountain is Everest's 29,140 feet—an elevation about thirteen miles above the floor of the deepest ocean trench. Everest's maximum rise above sea level would correspond to an elevation of about one one-hundredth of an inch on a globe thirteen inches in diameter.

In the perspective of astronauts, then, even the most rugged scenery is monotonously smooth. Only for the earthbound or the low-flying are there panoramas of strong relief and dramatic forms.

Landscapes in Time

Traditionally in geology, Earth's history has been traced by the study of the sequence of rock strata. The pioneer English geologist William Smith (1769-1839) discerned that rock strata in most occurrences become older as one digs deeper, and that the older the strata are, the more primitive will be the fossils—if any—discovered in them. Fossils thus are indicators of the relative ages of sedimentary strata. The fossils found in any particular stratum show approximately, and sometimes precisely, where this stratum belongs in the sequence of geologic history. Where fossil-bearing rocks can be dated, moreover, it may be possible to date likewise any nonfossiliferous rocks associated with them.

It is on this basis of relative dating, rather than dating in terms of years, that most geologic history has been reconstructed. Fossils and the rocks in which they are found are rich in clues not only to ancient life but to the physical events of the past—the rise of mountains and the spread of deserts, the waxing and waning of vast drainage systems, the invasions of continents by the sea, the periodic glaciations and the times of widespread volcanic outbursts. Working from region to region, geologists have pieced together a chronology for Earth as a whole, known as the Geologic Time Table. This outline is remarkably definitive from the earliest period during which living organisms left a readable record in the sedimentary rocks; but the previous history of Earth, perhaps 85 per cent of the total, remains obscure.

This chronology has four grand divisions, or eras, each ending with an episode of intensified crustal disturbances and abrupt changes in the character of animal and plant forms. The eras were named Archeozoic (Ancient Life) or Cryptozoic (Unknown), Paleozoic (Old), Mesozoic (Middle), and Cenozoic (Recent Life). Each era was divided into periods, which are separated from one another by changes less drastic than those separating the eras. The periods of the Paleozoic were given the names of localities where occurrences of the rock strata were studied by early geologists—for example, Cambrian for the region of Cambria in Wales, and Devonian for Devonshire in England. Periods of other eras were named otherwise—after mountains (Jurassic, from the Juras), abundant rock types of the period (Cretaceous, "chalky"), and so on. The Cenozoic Era, the most recent and most detailed in the surviving record, was divided not only into periods but into epochs as well.

The Geologic Time Table proved highly serviceable for generations before any reliable quantitative figures in terms of years were available for denoting geologic time. Since geologists agreed much more as to what the rock sequences are than as to the lengths of time they represent, it became customary to date rock in terms of the eras, periods, and epochs. In due course, however, absolute dating—that is, dating in terms of years—was attempted with increasing success. Around 1900 Earth's age was generally estimated at 100 million years—a figure derived from apparent erosion rates and the rate at which the oceans appeared to be receiving salts from the land via streams. This estimate was increased at intervals as closer studies of weathering and erosion were completed. Finally, in the 1950s,

A mountain stream at work: Rivers are the dissectors of Earth's crust. The mountains here are the ancient Grampians of Scotland at Glencoe.

GEOLOGICAL TIME CHART

Age in Years	Era	Period or Epoch	Important Physical Events
1.0 ±.5	CENOZOIC	PLEISTOCENE	Repeated extensions of ice caps in arctic and north temperate areas
			Continents generally elevated, mountains high, deserts widespread
13 ±1		PLIOCENE	Mountain building in NW North America
			Deformation of Tethys geosyncline; Alps and Himalayas rise
25 ±1		MIOCENE	Extensive erosion surfaces cut on Appalachians and Rockies
			Trend to cool, dry climates over much of the world
36 ±2		OLIGOCENE	Initiation of mountain building in Tethys geosyncline
			River and flood-plain deposits begin on Great Plains
58 ±2		EOCENE	Climates warm and uniform; widespread jungles and forests
65 ±2		PALEOCENE	Development of basins between ranges along Pacific Coast and Rocky Mts
135 ±5	MESOZOIC	CRETACEOUS	Mountain building in Rockies Seas invade much of W North America and cover Atlantic and Gulf coastal plains
180 ±5		JURASSIC	Widespread mild, uniform climates Mountain building along Pacific Coast of North America Extensive marine invasions of southern and central Europe
230 ±10		TRIASSIC	Fault basins in E North America Extensive deserts and dead seas in North America and Eurasia
280 ±10	PALEOZOIC	PERMIAN	Continents generally elevated Appalachian and Ural Mts complete their development Tethys geosyncline from Spain to India
310 ±10		PENNSYLVANIAN	Mountain building in S Appalachians and SW United States Ice age in southern continents Coal swamps in many parts of the world
345 ±10		MISSISSIPPIAN	Beginning of mountain building in S North America and central Europe Extensive seas over much of interior North America
405 ±10		DEVONIAN	Catskill delta built from New England mountains into New York and Pennsylvania Mountain building in NE North America Extensive submergence of geosynclines and interior of North America
425 ±10		SILURIAN	Formation of Caledonian Mts in NW Europe Dead seas in Michigan, New York, Ohio, SE Canada Deltas and gravel beaches along eastern edge of Appalachian geosyncline
500 ±10		ORDOVICIAN	Mountain building in NE North America Over 60% of North American continent covered by seas
?600		CAMBRIAN	Climates generally mild and uniform Seas invade North American continent Development of geosynclines around edge of North America
1,000 2,000 3,000 4,500		PRECAMBRIAN	Fault basins in Lake Superior region Deformation and mountain building through central North America Development of geosynclines throughout central North America Extensive mountain building in Lake Superior region Oldest dated rocks Probable origin of Earth from solar dust cloud

MILLIONS OF YEARS BEFORE THE PRESENT (ESTIMATES REVISED IN 1961) with the development of radioactive dating, the figure of 4700 million years became widely accepted.

Radioactive dating is a technique that can be applied to any rocks which contain a radioactive substance, such as uraninite. It is assumed that this substance was formed when the rock was formed, and that radioactive breakdown began immediately. Samples of the substance are analyzed to determine the percentage that has broken down to form the next element in the order of decay, and this percentage is then checked against the known rate of breakdown. Thus the age of the radioactive substance, and of the rock with which it originated, is determined.

Radioactive dating may be accurate to within about 1 per cent, and it has the advantage of being applicable to rocks, such as lavas, that lack fossils. Many rock formations that formerly yielded no clues as to age can now be dated with reasonable certainty. Accordingly, the Geologic Time Table has undergone certain changes. In particular, the hitherto poorly charted chronology before the Cambrian Period is appearing in more detail and is being extended in time.

For professional purposes geologists are usually content to date various landscape features according to the Geologic Time Table—as "late Cretaceous," for example, or as "early Pleistocene." Since landforms generally take shape and are eroded away over long time spans, efforts to date them precisely are likely to be pointless. Laymen particularly, however, like to know how old things are. If approximate dates can be mentioned, one can get a notion of how long certain familiar forms have been in the making, how long they are likely to last, and how they compare in the time dimension with other landforms.

Probably no portion of Earth's present surface is part of the original crust, although some lavas being erupted from oceanic volcanoes may be similar chemically to those of the planet's early days. The most ancient rocks—certainly those near the surface of the crust—must have been disintegrated by the elements long ago. Their remains have been scattered, buried, chemically altered and recombined to form new rocks—not once but many times. The oldest rocks known, discovered in Rhodesia, have been dated at 3600 million years, or about three fourths of Earth's estimated age; but these formations were originally sedimentaries made from the wastes of earlier rocks. The exposed very ancient cores of some mountain ranges, such as the Laurentians of Canada and the Bighorns of Wyoming, contain rocks with ages as great as 2400 million years. But most of the bedrock now exposed around the world consists of sedimentary strata with ages of less—usually much less—than a few hundred millions of years. The youngest rocks of the crust are the solidified lavas produced by volcanic eruptions of our time.

It is a certainty that none of our planet's original landforms—none of the original mountains and valleys, rivers and lakes, plains and seacoasts—survives. Of all these features mountain ranges have by far the greatest longevity, and this is due to the fact that over periods of tens of millions of years most ranges are uplifted not once but several times. Old ranges, furthermore, such as the Laurentians, today could bear little surface resemblance to the originals, because all the original outer rock layers are gone. In fact, much of the rock now exposed in old mountains was originally formed at great depths and has been uplifted and bared to the elements only recently. The particular shapes that give mountain ranges their individuality are, like the shapes of other landforms, the results of erosion, and perhaps crustal distortion and volcanism, that occurred only yesterday as geologic time is measured.

How old, in years, are some of the familiar landforms around us? Counting time from the original uplift, the younger (also called "Newer Folded") Appalachian Mountains of the eastern United States are about 250 million years old; but they have been cut low by erosion and uplifted anew several times. The Rockies, first uplifted about 60 million years ago, were all but leveled by erosion at least once and then uplifted again. The Alps, which began rising perhaps 30 million years ago and are still gaining, stand high despite the fierce pace of erosion on their slopes. The Himalayas, which are the youngest of the world's major ranges, have been undergoing uplift for about 20 million years and are still rising faster than erosion is destroying them.

The lowland occupied by the Mississippi River and its tributaries dates back scores of millions of years. Perhaps this lowland was formed originally by crustal distortion.

In contrast to the Mississippi Valley is the valley of the Colorado River along the stretch better known as Grand Canyon. This gorge can be said to date from the beginning of the recent uplift of the Colorado Plateau about two million years ago. The course of the river had become established much earlier, but it was the uplift that speeded the river's flow enough to cut the canyon.

Some 50 million years ago the general contour of the East Coast of the United States had become about what it is today. The Pacific Coast is much younger, having been built out seaward by intense volcanic activity 20 to 25 million years ago, and subjected more recently to severe crustal disturbances along its central and southern margins. But coasts are noticeably unstable, and even the Atlantic edges of the continent have been rising and sinking in some degree during recent time. Shorelines practically everywhere have been somewhat altered during the past ten to twenty thousand years by the rise of sea level due to the melting of the great glaciers of the Pleistocene. The details of shorelines—headlands and coves, beaches and bars—are features that change over spans of time ranging from thousands of years to months and weeks, or even days.

Rock, water, and time: Here in Grand Canyon of the Colorado are exposed layered rocks formed between the Precambrian Era and the Permian Period—a span of a billion years or so. Note the benches of strong Precambrian rock near the bottom of the gorge.

Volcanoes, even the largest ones, are usually young by geologic standards. They may be built up by eruptions of lava over a mere few centuries or millennia of activity. Many are all but destroyed sooner or later by especially violent outbursts. As their activity ceases, weathering and erosion quickly disintegrate and scatter the weak lava rock of the cone, and after a few thousands or tens of thousands of years little may remain of a once-imposing mountain. In the United States, volcanoes and their remnants are easily recognized only in areas of the West that have seen volcanic

A profile rock: The Old Man of the Mountain, a famous spectacle in Franconia Notch, New Hampshire, was shaped in granite by ages of weathering.

activity within recent time. Some well-preserved cones such as those of Craters of the Moon, in Idaho, and Capulin Mountain, in northeastern New Mexico, have ages ranging from five hundred to two thousand years. Mount Mazama in Oregon, the extinct volcano whose crater is occupied by Crater Lake, ceased activity about 6300 years ago. Ubehebe Crater in Death Valley, California, last erupted about one thousand years ago. The volcanic mountains of the Puy du Dome areas in the Auvergne of south-central France became extinct perhaps 2000 years ago. The eruptions that built up Italy's Mount Vesuvius are believed to have begun some ten thousand years ago. The entire Hawaiian Island chain is thought to be the product of a mere two million years of volcanism.

In geology, thousands and even millions of years "come cheap." It is obvious that all but the very youngest of landforms are, measured against human history, awesomely old. The uplift of the "young" Himalaya Mountains began ten to twenty times as long ago as the evolution of man from his humanoid ancestors. Grand Canyon, if we consider it to be one million years old (the rapid cutting began probably two million years ago), is two hundred times as old as the Pyramids of Egypt. The heaps of rock debris left on the northern United States by the melting ice sheets of the Pleistocene had lain there fifty centuries before the birth of Christ. Summing up, it is apparent that the major landforms of today were created long before man, have changed little during his history, and—barring the unimaginable—will look much the same after he has gone.

Toadstool rocks in the making: Odd features in the Painted Desert of Arizona, due to rainwash and natural sandblasting, appear in several stages of development.

III

Rocks, Air, and Water

In our solar system, Earth apparently is unique in having voluminous surface waters, a dense and chemically active atmosphere, a great diversity of chemically active minerals, and a range of temperatures that allows water to exist as vapor, liquid, and solid. These extraordinary conditions make Earth's surface a scene of pervading and unceasing change.

Atmospheric gases agitated by solar heating, by the gravitational pulls of Sun and Moon, and by Earth's rotation bathe all landscapes, slowly infiltrating and decomposing the rock. Plant roots, growing in the wastes of rock decay, penetrate rock crevices and widen them. Water freezing between grains and within cracks gradually breaks up rock masses. Changes in temperature weaken rock surfaces by causing them to contract and expand by turns. Winds wear back these surfaces by hurling dust and sand against them. Rain falling upon rock dissolves portions of it, and then, running downslope, carries along sediments that abrade it. Snows accumulating on mountains form glacial ice that, descending in ponderous masses, grinds out mountain valleys and scours lowlands. Gravity maintains an unrelenting pull everywhere, so that as rock fragments are loosened they fall, slide, or roll downward. Finally, rainwater and meltwater, forming sheets and streams, work with wind and gravity to keep the wastes of the land moving seaward.

Most of these destructive events involve water, and plenty of that is supplied through the hydrologic cycle. Air moving over lakes, rivers, and ocean picks up great quantities of water as vapor, then drifts over land and, when cooled enough, drops the moisture as snow or rain, which is thereupon variously evaporated into the air again, converted into ice, absorbed by rock and soil, taken in by plants and animals, drained into lakes and rivers, and returned to the sea again. Were the hydrologic cycle to end, so would the land-forming work that is done by water, and the crust would become almost changeless through future ages.

The attack of the elements on the crust produces what is called degradation—that is, the reduction of land toward sea level as rock is

Horizontal rock stratification: The influence of rock structure is emphatically clear in this deeply eroded terrain—Land's End, on the Colorado Plateau near the junction of the Green and Colorado rivers in Standing Rock Basin, southern Utah.

destroyed and the wastes are swept seaward. The process is continuous and unceasing. Highlands and lowlands alike are constantly losing their substance seaward. The continents would have vanished beneath the waves long ago were it not for the occasional periods of uplifting and upbuilding by diastrophic and volcanic events.

But downwasting is not the whole story. Degradation processes also carve and model most of the relief on the planet's surface—not only features due to downcutting and downwearing, such as the valleys that cut through highlands, but also relief formed by deposition—alluvial fans, flood plains, beaches. Crustal movements and igneous activity do produce some relief directly, such as block mountains and volcanoes, but even as these begin to take form, reshaping by the elements begins. Degradation processes are the dominant creators of relief on most terrains most of the time.

All degradational events are summed up in three terms: weathering, mass wasting, and erosion. Weathering includes the alteration and break-up of rock in place by chemical and physical processes, such as the freezing of moisture within rock or chemical reactions between air, water, and minerals to form new compounds. Mass wasting is the downslope movement of rock waste by gravity without the aid of running water or wind; rockfalls and landslides are obvious examples. Erosion, a term often loosely used to include weathering and mass wasting, means in its strict sense the destruction of rock by agents in motion—wind, water, glacial ice—and the transport of the debris to other places.

Rock Resistance and Rock Structure

The degradation process is always conditioned by the resistance of the rock. Resistance varies with the rock's chemical composition, its structural relationships, and local climate. Rock made up mainly of minerals that have a fairly stable chemical composition breaks down slowly and hence tends to stand higher than rock made of less stable minerals. Rock resistance thus accounts for the essential difference between highlands and lowlands on any landscape where the relief results primarily from degradation processes. But relief patterns may reflect also the sequences of the rock formations and the attitudes with which they meet the onslaught of the elements. Structure involves lines or bands of relative strength or weakness. It determines where the highlands and the lowlands will be, and may dictate whether relief profiles will be steep or gently sloping, rugged or flat or rolling.

One basic aspect of structure is jointing. Joints are fractures produced in bedrock by natural stresses and strains. All bedrocks have networks of joints, which may be closely or widely spaced, predominantly vertical or horizontal, or simply random. Since they allow easy entry to water and thus represent lines of weakness, they guide the shaping of much relief. Strong vertical jointing may cause the degradation processes to create rock towers, chimney rocks, needles, or finlike formations, such as one sees in Bryce Canyon, Utah, or the Dolomitic Alps of Italy. Jointing that is predominantly horizontal tends to produce wide, flat-topped features. Where joints are widely spaced, relief favors broad, rounded forms, and cliffs present bold, solid fronts, like those of the Sierra Nevada in California or the White Mountains of New Hampshire.

Horizontal rock strata may give a terrain a flat or rolling surface, as on the American Great Plains or the Appalachian Plateau. If the terrain has been strongly eroded, it may feature many isolated flat-topped hills, such as the mesas on the Colorado Plateau. Rock folds may produce a topography of long, parallel ridges like those of the Folded Appalachians. Where folds have been cut off by erosion, the exposed edges of the upturned strata may form ridges and valleys according to their relative resistance.

As erosion exposes masses of rock formed by intrusions of magma into the crust, these masses usually prove to be stronger or weaker than the

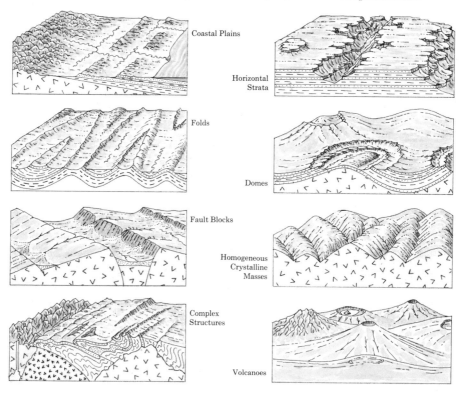

Erosion on various kinds of rock structures: Most landscapes, especially highlands, express the structural relationships of rocks. They express rock composition also.

adjacent rock and hence yield positive or negative relief. Thus parts of the Colorado Plateau are dotted with high-standing intrusive rock masses that are stronger than the surrounding sedimentaries, while the Adirondack and White Mountains exhibit many small valleys formed by the destruction of similar intrusive formations in highly resistant granitic rock.

Where crustal blocks have been uplifted, sunk, or tilted so as to make relief in the form of the cliffs called fault scarps, the displacements of rock masses of varying strengths beneath the surface will continue to guide the erosion process long after erosion has removed the original scarps.

Rock structure is expressed strongly when a terrain stands high and streams are swift and efficient in exploiting the zones of weakness. As land is reduced in elevation, the streams lose downcutting power and relief loses intensity; that is, the profiles begin to level out. Now the lines of weakness are becoming, as it were, blurred. With time, as the land mass nears base level and the streams grow sluggish, rock waste accumulating on the bedrock protects it from further destruction, and the influence of structure virtually ceases. Erosion continues, but is limited mostly to the transport of sediments seaward.

Rocks as Land-formers

An understanding of landforms must be based finally upon some knowledge of the chemical varieties of rocks. Most varieties are fairly consistent in their degree of resistance, and tend to occur in characteristic structures. The ability to recognize common rock types often is the key to an understanding of how a landscape has come to be what it is today.

The bedrock that makes up Earth's crust consists of combinations of 1500 to 2000 species of minerals—the exact number depending on how one defines a species. However, most rock consists of but a small number of species, and not more than about twenty rock types are common. With a little knowledge one can identify at least tentatively most of the rocks one encounters. A list of common rocks with incidental information about them appears in the accompanying table. Beyond this, some comments are in order concerning their land-forming characteristics.

Of the three broad categories of rock—igneous, sedimentary, and metamorphic—the igneous varieties are the only ones that might be called "original." These form from molten materials cooling within the crust or at the surface. Those that solidify within the crust are called intrusive, and those formed at the surface are extrusive. Although sedimentary rocks cover about 70 per cent of the continental areas, the igneous rocks form the bulk of the crust. Everywhere these once molten masses lie beneath other strata, and in many sectors they have broken through. Even the sediments on the sea bottoms are underlain by ancient lava rock.

Columnar basalt: Strains that develop in basaltic rock as it cools may fracture it to form long, six-sided (and sometimes five-sided) columns. This is the famous display at Devils Postpile, California.

The so-called basic igneous rocks, of which the commonest is basalt, apparently originate primarily from magma that has squeezed up from the mantle. Most intrusive and extrusive "basics" are hard and mechanically strong, but are rich in minerals that are relatively reactive with water. Under a humid climate, therefore, they offer weak resistance. On cooling, basaltic rocks may become closely jointed—that is, naturally fractured—

and thus are the most susceptible to destruction. Accordingly, in humid lands where granite or other highly durable rocks predominate, any basaltic masses that are exposed tend to become valley-formers. In the desert, because of the dryness, they frequently occur as relatively strong rocks, forming ridges, buttes, and protective caps on mesas.

In some volcanic regions, such as the Columbia Plateaus of the northwestern United States, basalt has been erupted in such quantities as to drown thousands of square miles of terrain. Most oceanic islands, including the Hawaiians, Iceland, the Canaries, and those off northern Scotland, were formed from basalts erupted from the sea bottoms.

The other main group of igneous rocks is made up of the acidic varieties. The common type is granite, light in color and not quite so heavy as basalt. Granites form at depth mostly from melted crustal rock and mantle material. Very resistant to water and often having joints that are widely spaced, granite is a maker of highlands. Most formations occur as the exposed cores of mountain ranges such as the Rockies and the Sierra Nevada, and as large domes from which covering strata have been eroded off, such as the Black Hills of South Dakota. Unless sculptured by valley

A rhyolite pinnacle at Chiricahua National Monument, Arizona: Lava dust was welded by heat to form rhyolite rock, then severely fractured by crustal movements, and finally eroded by rainwater streams following the fracture lines.

COMMON ROCKS*

	Name	Predominant Minerals	Origin	Description
Intrusive (plutonic) coarse-grained	GRANITE (acidic)	Orthoclase or microcline, much quartz, usually hornblende or mica	Solidification of magma at depth	Light-colored; crystals about same size; may be porphyritic (very large grains)
	SYENITE (acidic)	Orthoclase or microcline, kaolin		Similar to granite but smaller crystals
	DIORITE (basic)	Plagioclase, hornblende		Gray to dull green; grades into granite
	GABBRO (basic)	Plagioclase, augite, sometimes chlorite		Dark grayish or greenish; texture like granite
	PERIDOTITE (basic)	Olivine, pyroxene, amphibole		Nearly black; weathers greenish; heavy
Extrusive (volcanic) fine-grained	RHYOLITE (acidic)	Orthoclase or microcline, quartz	Solidification of lavas on surface	Light-colored; white to reddish or gray; usually fine texture but may be porphyritic
	OBSIDIAN (acidic)	Same as rhyolite		Noncrystalline, glassy, dark, sometimes transparent
	PUMICE (acidic)	Same as rhyolite		Very light, frothy
	ANDESITE (intermediate)	Plagioclase, augite		Often dark greenish; texture like rhyolite
	BASALT (basic)	Augite, plagioclase, often olivine		Dark, heavy, often greenish; very fine texture; weathers brown. Porous form is scoria.

IGNEOUS
solidification of magma or lava

	Rock	Composition	Origin	Characteristics
SEDIMENTARY consolidation of sediments	LIMESTONE	Calcium carbonate (calcite)	Precipitation of calcium carbonate in clear sea water; or shell accumulations	White to blue; massive, crystalline, or shelly; usually stratified clearly
	DOLOMITE	Calcium and magnesium carbonates	Replacement of calcium in limestone by magnesium	Darker than limestone
	SHALE	Quartz, other silicates (clay minerals)	Deposit of clays, especially by sluggish waters, on land or offshore	Flaky, sheety, or slabby; mostly red to brown, or other colors; fine-grained; often crumbly
	SANDSTONE	Quartz grains cemented by silica, calcium carbonate, iron oxide	Deposit of sand on deserts by wind, or more usually by streams near highlands or by waves and currents on shores	Joints usually rectangular; mostly red to brown; often very strong; breaks around (not through) grains
	CONGLOMERATE	Rounded rock fragments (often quartz) cemented frequently by silica, lime, or iron oxide	Deposit of sand and cobbles at foot of highlands or on beaches	Like sandstone but contains cobbles; joints irregular; often very strong
	BRECCIA	Angular rock fragments of composition depending on origin	Consolidation of any large-caliber rock debris except cobbles	May be like conglomerate, but properties depend on origin
METAMORPHIC alteration by heat, pressure, and infiltration	GNEISS	Quartz, orthoclase or microcline, mica; or hornblende, orthoclase or microcline, quartz	Metamorphism of shale, sandy shale, or igneous rocks	Often has banded or pepper-and-salt aspect; blocky or irregular jointing; many colors; usually very strong
	SCHIST	Mica, quartz, orthoclase or microcline, or chlorite, epidote, quartz, plagioclase; or hornblende, plagioclase, quartz	Same as gneiss but with less intense metamorphism	Wavy or contorted lineations; may resemble shale, but stronger; flaky
	SLATE	Mica, quartz	Metamorphism of shale	Gray, green, or red; breaks across bedding planes; may be folded or wrinkled; usually strong
	QUARTZITE	Quartz; or quartz, orthoclase or microcline, kaolin	Metamorphism of sandstone	May suggest sandstone but breaks through (not around) grains; can be polished; very dense and tough
	MARBLE	Calcite	Metamorphism of limestone or dolomite	White grading into other colors; often intricately streaked; dense and hard unless crystals are large

* Handbooks to aid in the identification of rocks and minerals are listed in the Bibliography, page 349

A land blanketed with lava dust: Arizona's Painted Desert has been shaped in deep beds of varicolored dust blown out by now-extinct volcanoes. A solidified lava flow (*upper right*) protects the soft underlying deposits from erosion.

glaciers, as in the Sierra, granite uplands are likely to display rounded summits and steep sides, like those of the Hudson Highlands in New York state.

Rhyolite is a granite-related rock that forms from lavas erupted by volcanoes. Generally it is not so resistant as granite. Layered rhyolite flows sometimes become eroded into irregular pillars like those of Chiricahua National Monument in Arizona and the Cimarron Valley in northeastern New Mexico.

Intermediate between basaltic and granitic rocks is the andesitic type, which forms from mixed basic and acidic materials. Andesites are erupted along the volcanic "ring of fire" around the Pacific Ocean.

Some magmas when erupted form a very gassy lava. As the gases

ROCKS, AIR, AND WATER · 41

escape, they leave the cooling mass highly porous; thus such rocks as scoria and pumice are created. These make up portions of volcanic cones and may be much fragmented and widely scattered by explosive eruptions.

The commonest rocks at the planet's surface are the sedimentaries. These originate from rock waste, chemical precipitates, and organic remains accumulated on lowlands, on river and lake bottoms, and on sea bottoms offshore. The sedimentaries are stratified rocks of many colors, deposited mostly horizontally but often becoming distorted, tilted, or displaced later. Shales, sandstones, conglomerates, and limestones make up the chief groups. As already mentioned, they cover about 70 per cent of the continental areas, mostly on plains, plateaus, and the flanks of mountains, and they lie as a veneer on much of the ocean bottom.

The commonest sedimentaries are the shales and sandstones. These lie at or near the surface in most of the continental United States and generally over Canada except in the southeast. They derive from well-weathered materials and hence are little affected by chemical weathering. They resist erosion poorly unless the grains have been cemented firmly by infiltrating mineral solutions. Shale, which is sheety or slabby in structure and often mechanically weak, is usually a valley-former. Sandstone is likely to have a blocky aspect because of rectangular jointing; often it is strong enough to stand well above other sedimentaries.

Conglomerate resembles sandstone except that it is made up partly of large rock fragments rounded by stream or wave action. A similar but less

Shale outcrop: These strata in southeastern New York formed from black muds accumulated on a delta about 400 million years ago. Later they were folded and uplifted. This outcrop is the stump of a fold cut down by weathering and erosion.

A sandstone exposure: This is part of the surface of Checkerboard Mesa in Zion National Park, Utah. The rock formed from sands deposited by ancient winds, whose changes of direction and speed produced the crossbedding visible in the photograph.

common rock, breccia, consists of cemented fragments of any rock type that have not been water-worn and thus are angular.

Limestones make up the surface formations in the chain of lowlands called the Great Valley, east of the Folded Appalachians from New York to Tennessee. Covering nearly all of Florida, they are exposed widely also over the eastern and midwestern states between Wisconsin and New York in the north and the Gulf of Mexico in the south. They crop out in sectors of the West, including the northern Rockies, western Texas, the Bryce and Grand Canyon localities, and some highlands of New Mexico.

Consisting mainly of calcium carbonate, limestones may be made up chiefly of either shells, fine grains, or crystals, poorly to well consolidated. A variant of limestone is dolomite, in which much of the calcium in the carbonate has been replaced by magnesium, which darkens the rock. Limestone and dolomite are of special interest because of their easy solubility in water that has absorbed carbon dioxide from the atmosphere. Under a humid climate the rock is removed relatively fast by rainwater and thus

Coarse conglomerate: This rock of southeastern New York formed from rock debris dropped by streams at the foot of a highland. Only a stream on a steep slope could have moved cobbles of the size (up to a foot long) seen in this rock.

typically is found on lowlands. Here and there, special conditions have left a formation high enough so that water drains through it rapidly, dissolving out sinkholes and networks of tunnels and caverns, and perhaps cutting deep gorges. On arid terrains the carbonate rocks often show greater strength than sandstone and shale, and accordingly may stand as highlands.

The third major rock group, the metamorphics, include rocks of any kind that have been metamorphosed—physically and chemically changed —by heat, pressure, or infiltrating fluids at depth. Especially common are schist, slate, and gneiss. All these can derive from shale, but certain schists and gneisses also come from igneous rocks. Quartzite is metamorphosed sandstone, and marble derives from limestone. Slate, gneiss, quartzite, and—less often—schist may be highly resistant to both weathering and erosion, and hence they are often highland-formers.

Quartzite, perhaps the strongest of all rocks, tends to make round-topped hills with cliffy sides, as in the Baraboo Range of Wisconsin. Usually

Chalky limestone: In west-central Alabama the Selma chalk, consisting of the limy remains of minute marine organisms, is being cut through by the Tombigee River.

it weathers off in rectangular blocks. By contrast, schist and slate strata, especially where they have been sharply tilted and erosion is occurring on their upturned edges, can make sharp-backed, ragged highlands, as in the southern Appalachians and the Taconic region of eastern New York.

Gneisses, often showing bands of light and dark minerals, are likely to be widely jointed and massive. Often they rear above surrounding sedimentaries to make rounded uplands such as the Ramapo Mountains of New Jersey, the Hudson Highlands of New York, and the Great Smokies of North Carolina and Tennessee.

Coarse-grained marbles are weak. Others are massive, irregularly jointed, and mechanically strong, but being vulnerable to water, they make lowlands where the climate is humid. Some valleys of Vermont's Green Mountains are in marble.

Metamorphic rocks are found typically in mountainous regions where the crust has been under great stress and strain, and has been invaded by hot fluids. Belts of metamorphics often are found flanking mountain cores that are of granitic rock, as in the Adirondacks and the White Mountains.

Schist: This exposure on Pico Peak in southern Vermont shows intense deformation. The whitish material is intruded quartz.

Banded gneiss: The rock of this exposure in southern Vermont was infiltrated with hot fluids that formed bands of quartz. Banding may follow original bedding planes.

Jointing in granite: This rock in the Hudson Highlands, New York, is well advanced toward destruction by weathering.

Jointing in basalt: Progressive cooling and solidifying of a lava flow from the surface inward produced a concentric pattern of joints. This outcrop is in a road cut at Great Notch, New Jersey.

Exfoliation of granite: The breaking-off of large shells on Popolopen Torne, in the Hudson Highlands, is due probably to expansion following removal of the weight of covering rock.

Weathering by Mechanical Forces

The atmosphere's work on rock is slow, but ubiquitous and ultimately triumphant. In the perspective of geologic time, the loosening and fall of each rock grain is momentous. We look at the hills to see where the grains are falling, and they are falling everywhere.

Most of the bedrock that is now bared to the elements was formed long ago beneath other rock that has since eroded away. As erosion has lightened the load on buried rocks, they have expanded. One result has been the formation of systems of joints. Another has been the breaking off of the rock at the surface as flakes or shells (the phenomenon known as exfoliation) or as separate grains. These "unloading" effects are very pronounced in dense crystalline rocks such as basalt, quartzite, and granite.

Exfoliation was long believed to be caused mainly by temperature changes. When a rock mass warmed by a day of sunshine is chilled by rain or the coming of night, the surface contracts a little around the still-warm interior. Strains thus produced may crack the surface. Desert travelers have reported hearing cracking and popping sounds coming from rocks at nightfall. But exfoliation caused in this way is now thought to be uncommon. Laboratory tests indicate that temperature changes are more likely to produce grain-by-grain disintegration than exfoliation. Grain-by-grain surface disintegration does tend to produce rounded surfaces on homogeneous rock because protuberances, having a large surface area relative to volume, weather faster than the main rock mass. This process may account for the rounding of rock fragments that have been long buried in soil and thus protected against sudden temperature changes.

Boulders of exfoliation: Devil's Marbles in Northern Territory, Australia, are of granite, with diameters up to 20 feet. Note exfoliation domes in the background.

Weathering into rounded forms is typical of homogeneous rock with widely spaced joints. Wide jointing leaves the minimum surface exposed to weathering, and it limits the prying out of blocks by freezing water and tree roots. Disintegration thus occurs mainly by exfoliation or grain by grain, and it is so slow, particularly if granite is involved, as to allow protuberances to be eliminated.

Exfoliation is common on the exposed gneiss and granite cores of ancient highlands such as the Laurentians of Canada, the Hudson Highlands of New York, and the Great Smokies. Rounded granites are familiar in the Black Hills, but here one sees also the unusual sight of granite "needles," which result from strong vertical jointing. Probably the most famous exfoliation domes in the United States are Half Dome, in Yosemite National Park, and Stone Mountain, the well-known isolated granite mountain in Georgia.

Rock expansion is caused not only by unloading but by crystal growth. As is well known, the volume occupied by water increases by about 10 per cent when it freezes, and if the water is confined the expansive force is tremendous—some two thousand pounds per square inch. Wherever water penetrates between rock grains or into joints and freezes there, destruction occurs. Even in the warm tropics, in fact, crystallization can be a strong weathering process, for here as elsewhere minerals such as halite (common salt) and quartz may be precipitated from water percolating through rock, and the precipitates can crystallize expansively much as freezing water does. As might be expected, crystal growth tends to produce grain-by-grain disintegration if it occurs on rock surfaces, and jointing or exfoliation if it occurs within the rock.

Rock destruction by freezing is variously called frost wedging, frost prying, or ice prying. It produces thick rock litter in cold lands where rain and snow are abundant and the thawing-freezing cycle occurs during much

Felsenmeer: Large rock fragments from frost-prying accumulate on the broad summits of the White Mountains of New Hampshire. Most small-caliber rock waste has been blown or washed downslope.

Destruction of rock by plants: Constant pressure exerted within rock joints by the growth of tree roots widens the joints and accelerates weathering.

of the year. On the slopes of polar highlands and of mountains in temperate regions one finds impressive accumulations of rock fragments testifying to this cycle. Especially vulnerable are rocks that are much jointed, highly porous, or weakly consolidated along the bedding planes.

Above timberline on mountain slopes one often comes to the kind of boulder field known as felsenmeer (German, "sea of rock"). This litter of ice-shattered rock consists mostly of larger fragments, because the small ones have been blown away by strong mountain winds, washed downslope by rain or meltwater, or partly covered by alpine plants. Felsenmeers can be looked for on mountains in North America as far south as the White Mountains in New Hampshire, the southern Rockies, and the Sierra Nevada. The higher levels of Pikes Peak, in Colorado, offer a fine display, as does the Presidential Range in the White Mountains. Felsenmeers are to be distinguished from boulder litter left on lowlands and highlands alike by the Pleistocene ice sheets.

In humid regions of the temperate and tropical zones, especially, one of the most vigorous weathering agents is ordinary vegetation. Plant roots feel their way into rock joints and, growing within them, exert a slow but powerful expansive force. Everyone has seen sections of sidewalk tilted, raised, or cracked by the growth of tree roots; and so it is with rock. On humid hillsides the birch seems especially adept at finding rootholds in rock crevices and growing to a size that can cause tremendous strains upon the rock. Where the rock is too massive to be split by roots, these may nevertheless prepare the way for other agents of destruction—for example, by holding water that may freeze.

Weathering by Chemical Agents

In chemical weathering the dominant processes are hydration, oxidation, and carbonation. Hydration and oxidation break down common silicate minerals—feldspars, pyroxenes, amphiboles—to form the clay minerals that make up the bulk of soils. Another mineral, quartz, is left unchanged by these processes, but minute quantities can be picked up and carried by running water in colloidal suspension, and over long periods these minute quantities bulk large. The basic rocks are much more vulnerable to chemical destruction than the acidic rocks and those sedimentaries that have formed from already-weathered rock wastes.

Selective chemical weathering is demonstrated by the so-called solution hollows often seen on exposures of granite or gneiss. On Bear Mountain, New York, the granite consists of light acidic minerals with scattered aggregations of dark basic materials. The acidic material, which contains much quartz, strongly resists attack by moisture, but the basic minerals become chemically altered and are dissolved out. The resulting depressions trap water and thus accelerate disintegration.

The carbonation process starts with the union of water and carbon

Solution holes in granite: Concentrations of the relatively weak dark minerals weather out faster than the light minerals, leaving depressions. These are on Bear Mountain, New York.

dioxide from the atmosphere to form carbonic acid. This weak acid reacts with insoluble calcium carbonate in limestone to form the soluble bicarbonate, which is then carried off by running water. This solution process accounts for the oddly beautiful scenery of limestone landscapes such as those in Indiana, Kentucky, and Virginia. Carbonation occurs also as a result of attack by carbonic acid on certain silicates in basalt and various gneisses, but these rocks are far less vulnerable than limestone.

Living things, too, participate in chemical weathering. Animals and plants by respiration and decay contribute to the atmosphere's content of carbon dioxide. Humic acids in organic wastes attack rock surfaces directly. Remains of dead plants and animals accumulate in the soil, and ground water picks up the acids of decay and carries them down to bedrock.

Solid rock looks waterproof, but water can penetrate it to depths of hundreds of feet, traveling by way of joints and the interstices that exist between rock grains or crystals. Some rock is so dense that penetration can be accomplished only by particles of molecular size. Water molecules are plentiful, however, and so is time.

As temperature and humidity increase, chemical weathering speeds up. In warm, wet regions, therefore, the reduction of rock to soil is rapid.

Camel Rock: This oddity north of Santa Fe, New Mexico, results from destruction of weak strata underlying stronger rock.

Water percolating downward through the soil picks up the soluble minerals easily and carries them seaward. Here one sees occurring at its quickest pace the worldwide leaching of minerals from the land by running water—the process that answers the old question of why the sea is salt.

Mechanical and chemical weathering support one another. The loosening of rock grains by temperature changes renders them more vulnerable to chemical attack. Hydration and oxidation may increase the volume of rock and thus set up internal strains that result in fracturing, exfoliation, or the shedding of grains. The new minerals formed by chemical reactions are sometimes porous and crumbly compared to the minerals from which they are derived, and thus may disintegrate rapidly under the assault of the elements.

Rates of chemical weathering, particularly, are suggested by surface deterioration on stone buildings and monuments. London's air has so damaged the limestone of Westminster Abbey that extensive reconstruction has been necessary—for safety as well as for appearances. The damage has been done mostly during the past century by the combination of moist air with gases from factory smoke and the exhausts of internal-combustion engines. In New England the graveyard tourist finds sandstone markers of the colonial period that have lost their inscriptions entirely, slate markers that are readable after three centuries, and marble and granite stones that show only slight attrition after a century. Going deeper into the past, there are the marbles and granites of ancient Egypt, Greece, and Rome, pitted and hollowed but still essentially intact after twenty centuries or more. These owe their condition mostly to a dryer climate than exists in England.

Mass Wasting: Attack by Gravity

Land looks static most of the time, but if one can visualize a mountain with all its myriad surface changes speeded up, so that a million years are concentrated into a few minutes, then the scene is static no more. The mountain's summit and the rock outcrops on its slopes are shedding blocks and flakes rapidly. Water is dashing furiously down all the gullies and ravines. Crashing landslides occur every few seconds. The soil cover, complete with trees, underbrush, and embedded rock slabs and boulders, is visibly flowing downward. Gravity is keeping the entire surface of the mountain in motion and is, quite plainly, shrinking it and leveling it.

So it is on the highlands of reality, except for the different time scale. The process of mass wasting is powerful and unceasing. As each rock fragment is detached by weathering, gravity begins pulling it downward. Individually and in masses, rock debris moves lower by falling, rolling, sliding, and even flowing. At intervals it is moved also by the agencies of erosion—wind, rainwash, streams, perhaps glacial ice. Any particular rock fragment travels only intermittently and is moved by different agents at different times, but gravity never forgets it.

Some rock debris suffers practically complete disintegration as it descends and some is held for long periods in talus slopes, valley fillings, desert dunes, or river deltas. Some wastes become involved in the formation of new sedimentary rocks. But the ultimate goal of all of it is the bottom of the sea, and some does get there.

The effectiveness of gravity obviously depends upon slope. A bit of rock detached by weathering on a hill will descend as far as the slope is adequate to keep it moving. Brought to a halt by friction or by some blocking object, it will start moving again as soon as anything happens to dislodge it. Small fragments can be started moving again by the impact of a raindrop, the touch of a bird's foot, or even a change in temperature. Large fragments remain immobile for centuries, but in time are certain to be set in motion again by frost-wedging or undermining or perhaps an earth tremor.

On a gentle slope most rock debris is so long in descending that it disintegrates rather thoroughly before reaching the bottom. Hence the relative scarcity of boulders and cobbles in a broad valley—unless such material has been dumped there by a glacier. On steep slopes the material moves faster, usually forming at the bottom of the incline the familiar talus slope

Rock creep: Blocks originally forced apart by ice-prying are being gradually pulled downslope by gravity.

ROCK, TIME, AND LANDFORMS · 54

of large fragments. These accumulations, like aprons at the bottoms of cliffs, are most noticeable in arid and polar lands, where they are not likely to be hidden by vegetation. If the rock is strong and widely jointed, as granite, basalt, and some other rocks may be, the talus will contain many large blocks. Sheety, weak rocks such as shale will produce sheety, much-broken talus.

A talus of simple rock fragments will generally hold an angle of about 35 degrees. A steeper slope tends to be unstable, and additions make it wider rather than higher. Where the debris accumulates at a high rate relative to the speed of its removal by water and wind, as often happens on deserts, the land may appear to be "drowning" in waste. A tourist sees excellent examples of such scenery while driving among old highlands of the Mohave and Sonoran deserts in the southwestern United States, or in deserts of North Africa and the Near East.

Deep talus, especially if it contains clay, may become internally lubri-

Effects of wide jointing: Strong homogeneous rocks, such as the granite gneiss here, often are widely jointed. The wider the jointing, the larger the size of the blocks that are weathered off.

Rockfalls caused by an earthquake: When the land shakes, the normal downward journey of rock waste on highland slopes is accelerated. Rockfalls here were caused by the earthquake in West Yellowstone in 1959.

cated by melting snow and ice. Then a steep slope cannot be maintained, and the mass begins to flow. Such a flow, called a rock stream or a rock glacier, is seen usually on cold-climate terrain where stabilizing vegetation is sparse. Some rock glaciers of the western United States, now "stalled," apparently are relics of the recent glacial ages. A number of these are seen in the vicinity of Silverton, Colorado. In Europe they occur in the Engadine region of the Alps.

On moderate slopes in humid regions vegetation slows the descent of rock waste and provides more time for it to disintegrate on the way down. Rock grains mixed with plant and animal remains form soil for more plants to grow in; and so the bedrock of the slope becomes covered with soil, roots, and some larger rock fragments. This mantle material may appear stable but over a long period does move. Each grain is disturbed intermittently by rainfall, plant growth, frost heaving, temperature changes, and chemical action; and each time it is disturbed it is likely to descend slightly before coming to rest again. In the same way, when a mass of rock debris freezes it expands somewhat perpendicular to the slope, and when it thaws it drops a little vertically, and so moves downslope also. By such move-

Rock fragments sorted by frost: The scene is in an area of patterned ground near Thule, Greenland.

ments there occurs a steady downward creep of the entire weathered mantle.

Some slopes exhibit a stepped profile due to slumping. Mantle is said to slump when it is undermined and sags, with noticeable backward rotation. Slumping may occur by stages, each of which makes a step in the profile. Common agents of undermining include waves, the flow of rivers, and man. Slumping is familiar in steep embankments along highways, especially where vegetation cover is thin.

Downslope movements of particles or clusters of particles separately are called soil creep. The net movement may be noticeable only over periods of years. When, however, rock debris becomes saturated or nearly saturated with water, the movement becomes a flow—that is, a mass movement. This kind of flow, common in alpine regions and at high altitudes elsewhere, is known as solifluction.

The results of freeze and thaw are striking in the "patterned grounds" of cold terrains. These are rock or soil masses formed into nets or polygons, rings, garlands, and stripes. Exactly how these figures are formed is not known, but the basic cause appears to be repeated freezing and thawing of concentrations of water along with variations in textures of the materials and the continuing action of gravity. These features can be found not only in polar regions but also on mountains in the temperate zone—for example, the higher White Mountains of New Hampshire.

Slides in the Adirondacks: Mount Colden (*at right center*) is noted for the breadth of its bared surfaces. Wide jointing of the strong anorthosite bedrock prevents vegetation from getting securely established.

Earthflows are movements that occur especially in water-saturated clay and silt. They resemble slumping, but there is no backward rotation of the material. The movement is faster than soil creep, but noticeable only after hours or days. Many forested slopes show scars—expanses of bared rock—left by these flows. Some such scars, called slides in the Appalachians and other Eastern highlands, are due to full-scale landsliding rather than to slow flow.

In arid country the mudflow is notorious. Usually developing with a torrential rain, it follows a definite channel, in which it may move at a speed of one mile per hour or faster. Mudflows from highlands in the Southwest have inundated roads and even buried small mining towns.

Even more dangerous are the mudflows started on the slopes of volcanoes when masses of volcanic ash become saturated with rainwater or condensed steam from an eruption. It was such a flow that buried the town of Herculaneum during the historic outburst of Vesuvius in A.D. 79.

Perhaps the best-known mudflow in North America—a harmless one, as far as man was concerned—was the tremendous "Slumgullion" movement. A broad river of saturated volcanic ash flowed down from the San Juan Mountains of Colorado, blocking the Lake Fork of the Gunnison River and forming scenic Lake Cristobal. The flow was nearly six miles long, and it left in the valley a deposit of mud over two hundred feet deep. The date of the flow is not known, but it apparently occurred within recent centuries.

Talus on mountain terrain: In the Beartooth Mountains, Montana, frost-broken rock tumbles down the slopes. Note the landslide debris to the left of the lake.

Swift mass movements of rock and soil down steep slopes are known as landslides. The usual causes are the weakening of soil and rock structures by weathering, excessive water content, or undermining by streams. The likelihood of slides is increased if the rock strata dip toward the base of the slope or are slippery when wet, as shale generally is. These two conditions are causes of frequent landsliding on the southern coast of California. Slides are favored also if the rock is much jointed, or if the freezing-thawing cycle is active, or if rains are frequent and heavy. Vegetation cover will tend to stabilize the mantle, but once a slide has begun, the largest trees may go with it. At the foot of very steep slopes one sometimes sees a stairlike or terrace pattern produced by successive slides during the recent past.

Heavy, rapid downslope movements of snow or ice are called avalanches. These can be started by such disturbances as thawing, strong wind, or blasting. Avalanches are especially hazardous to mountain climbers and skiers, who have attributed some of them to such slight causes as a footfall or a shout. In alpine regions the dreaded ice avalanche occurs when thawing detaches the lower end of a glacier and sends it roaring down the valley. Such was the ice avalanche from Mount Huascarán in Peru which, a few years ago, plunged downslope at an estimated 65 miles per hour, traveling ten miles, burying eight villages, and causing the deaths of over 4000 people.

Erosion: Master Sculptor

In erosion, several agents play interrelated roles. Sediment-bearing winds abrade rock surfaces, particularly in deserts, and transport the wastes elsewhere. Glacial ice scours, plucks, and gouges the land, and carries the debris down to the lowlands. Waves and currents attack shores with techniques of abrasion, hydraulic wedging and ramming, suction, and solution, and they distribute the sediments along the coasts. Finally and predominantly, running water abrades, dissolves, rams, sucks, and wedges the rock,

Desert stream work: On steep slopes of weak rock, deep and closely spaced gullies are typical. This erosion scene is in Death Valley, California.

Rock destruction in high gear: Mountains in the vicinity of Lake Louise, Alberta, show strongly the consequences of glaciation and frost-prying.

and carries off most of the wastes produced by all degradational agents together.

All these processes are so powerful that a book about landforms becomes a book mostly about erosion. Separate chapters must be given to the activities of rivers, wind, ice, and waves and currents. Running water, however, is so overwhelmingly dominant almost everywhere—even in the desert—that it demands attention before all other agents.

Water moves over the land as sheets and as streams. Rain falling on rock surfaces makes thin sheets which, finding the steeper gradients, usually form rivulets that run into gullies, which then lead the water to brooks, and these in turn run into rivers. Occasionally a rain is heavy enough to produce a sheetflood on a slope where gullies are widely spaced or shallow. Sheets may form also where torrential highland streams fan out as they reach a valley floor.

The potency of these sheets and streams is suggested by the amount of water that falls upon landscapes in the course of geologic time. Where rain averages 40 inches per year, as in the vicinity of New York City, about 25 gallons falls annually upon each square foot of area. This is at the rate of about 25 million gallons per square foot per million years. But water does not stop where it first strikes: it keeps moving, dragging its abrasive load for miles, or hundreds of miles, or even thousands.

Raindrops strike lightly but erosively. The impacts dislodge rock particles already loosened, and new surfaces are thus exposed. After impact the water, seeking lower levels, begins its abrasive and solutional work. As it gains in volume and load, its destructiveness increases geometrically. Finally, on reaching a valley, the stream loses gradient and starts depositing the large-caliber sediments. The bedrock may thus receive a protective cover, but erosion does not completely cease. Bit by bit—and in large quantities at flood time—the water keeps the sediments moving downgrade. As long as water is above sea level, it retains potential energy for erosion.

The valley-forming works of great rivers distract us from appreciating the accomplishments of small streams. Rivulets and thin sheets form on every landscape with every rain and every melting—including terrains on which no permanent streams exist. Much work is done by water that is evaporated before reaching any river. Spectacular examples of the erosional work of temporary little streams, with some help from weathering and mass wasting, are seen in the sharp, intricate relief of sites like Bryce Canyon, Utah, and The Needles in the Harney Peak region of South Dakota's Black Hills. It is the little temporary streams that dominate in the shaping of most highland profiles above the origins of the permanent brooks. Every landscape testifies to the work of a million nameless rivulets.

But much erosion occurs underground, too. It is estimated that the crust contains some two million cubic miles of water, not counting what is in lakes, streams, and ocean, and this underground water especially near the surface is constantly moving because of the pull of gravity and intermittent additions by precipitation. Like surface streams, it dissolves rock materials and carries them seaward. Underground solution and drainage account for the formation of caves and tunnels such as are found at Mammoth Cave and Carlsbad Caverns. Formerly thought to be of limited effect, underground erosion is now known to remove enough material from the crust—even in zones of poorly soluble rock—to make pronounced differences in our landscapes.

Where weathering, mass wasting, and erosion are occurring together, they may be almost indistinguishable. The dissolving of minerals from rock during rain may constitute either weathering or erosion, depending upon hairline distinctions. The tumbling of a rock fragment downslope in a sheet-flood involves both erosion and mass wasting. Erosion, however, is always supreme. It is erosion that keeps exposing new surfaces to attack by weathering and that does nearly all downcutting. River valleys, as we shall see, acquire their depth by stream work, and only as a result of this trenching process are there any valley walls for weathering and mass wasting to wear back. And it is by way of river valleys that the wastes of degradation are disposed of.

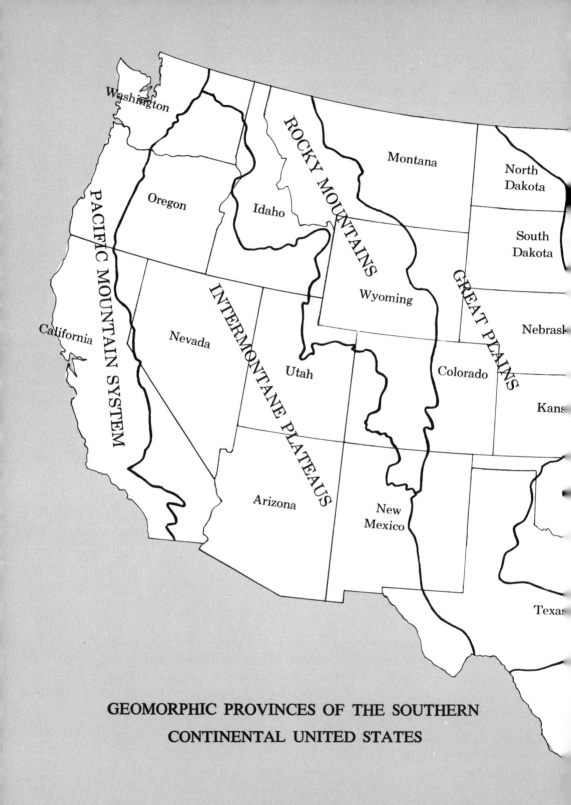

GEOMORPHIC PROVINCES OF THE SOUTHERN CONTINENTAL UNITED STATES

IV

Riverscapes

To a casual observer on an upland, in a valley, or on a plain, the relationship of a river to the surroundings is hardly more apparent than the fact that the world is round. In parts of Grand Canyon, which is the classic model of river work, the river is but a thread in a gaping valley. Even at high water there is little evidence that the channel is being cut deeper, the valley walls retreating. On wide terrains of low relief the role of the river as sculptor is difficult to visualize at all, because here the erosion cycle is nearly done and the river's work now consists mostly of the transportation and deposition of erosion debris. It is small wonder that among the ancients, with their much-abbreviated view of geologic time and change, the work of streams was grossly underestimated.

Around the turn of the eighteenth century educated men still believed that rivers began in the mountains at sites where water from Earth's interior, forced upward by the heat, burst through the crust. The waters were supposed to flow down through ready-made valleys until—unless lost by

The Connecticut River in old age: Here near South Deerfield, Massachusetts, the river winds southward toward its water gap in the Holyoke Range, which is visible on the horizon.

Drainage patterns of North America: A simple map of major watercourses follows the relief of the continent and suggests the importance of stream erosion in land-forming processes.

leakage through the valley bottom and sides—they reached the sea, from which they traveled via tunnels up into the mountains again. Later, snow and ground water became recognized as important sources of river water. But it was nearly 1800 before observers like James Hutton began to grasp the full nature of drainage and its commanding role in the shaping of relief. Today, looking down on landscapes from an airliner, we see clearly what was so difficult for our earthbound ancestors to see: the systematic, efficient patterns of erosion, so obviously the work of running water. Had the people of the seventeenth and eighteenth centuries traveled by air, almost certainly their geologizing would have been far more accurate.

Dissecting the Continents

Rivers represent the power of running water at its greatest, and their activities prepare the way for other forces of destruction. So important are rivers as the basic land-formers that geologists often describe entire regions in terms of their drainage and shaping by river action.

Relief almost everywhere clearly represents drainage. Gullies and ravines, as we have seen, result mainly from stream action. It is no accident that most valleys are occupied by rivers, either permanent or temporary, and it can be demonstrated that nearly all these valleys were cut by the rivers themselves. Only rarely does a stream occupy a ready-made natural valley. Indeed, most valleys that are now dry were cut ages ago by streams that have since disappeared because of climatic change or diversion of the water supply. Even inland areas that are today flat and riverless were at one time cut up by streams.

Every continent has far-flung drainage systems, in each of which one master river receives water from multitudes of tributaries. Each of these is supplied by smaller streams, which are nourished by still smaller ones, ad infinitum. The master rivers are well exemplified by South America's Amazon, Europe's Rhine, Asia's Yangtse, and North America's Mississippi. Drainage to these rivers has shaped the relief of millions upon millions of square miles.

The Mississippi with its tributaries drains 1,243,000 square miles, or about a third of the continental United States. Annually it pours into the Gulf of Mexico some 135 cubic miles of water—enough to make an inland sea 25 miles long, 5 miles wide, and more than a mile deep. This represents an average flow of 620,000 cubic feet per second and makes up 34 per cent of all the water discharged into the oceans by rivers in states south of Canada. Moreover, the quantity of sediments poured by the Mississippi into the Gulf is over 500 million tons per year, or 1,400,000 tons per day, some 90 per cent of this being in suspension. All this is material that has been torn, scraped, or dissolved out of the land.

Debris of a continent: Rock waste from a 1,243,000-square-mile drainage system forms the Mississippi Delta. This is Dennis Pass, near the river's mouth.

This destruction has been going on since early Cenozoic times, perhaps 40 to 50 million years ago, when the present Mississippi Valley was evolving toward its modern form. Drilling and seismic sounding along the coast of Louisiana indicate that the sediments there are thirty thousand feet deep, or nearly six miles. The state of Louisiana is in fact the site of an old basin which the Mississippi has filled with sediments. Today the river is working more leisurely than it did just after the Pleistocene, when it was carrying enormous volumes of glacial meltwaters, but the land in the drainage basin is still being reduced, the rate being about a foot every five thousand years. Considering the portion of North America that stands above sea level, the loss of a foot of elevation every fifty centuries may seem petty, but again the significant fact is the length of geologic time. One foot per five thousand years corresponds to one mile in 50 million years—not a long time span as Earth's history goes.

Compared to the Amazon, the Mississippi is modest. The Amazon's rate of discharge, by far the world's greatest, is estimated at 770 cubic miles per year. Next after the Amazon is the mighty Congo, running a poor second with its 340 cubic miles. America's Colorado River, considered by usual standards to be a river of respectable size, discharges only about 5 cubic miles of water annually.

All the world's rivers together discharge about 8400 cubic miles of water into the seas annually, or about 23 cubic miles daily. Water in all streams at any given moment approximates 300 cubic miles, equivalent to some thirteen days of flow. The latter figure appears perilously inadequate but is actually somewhat misleading, because most rivers are replenished continuously by the inflow of ground water. The reservoir of ground water in the crust is believed to be about two million cubic miles, or nearly seven thousand times the content of streams.

Only the water cycle, of course, keeps the rivers going. The oceans contain some 317 million cubic miles of water. From them the atmosphere picks up molecules of pure water at the rate of about 29 inches of ocean depth per year. Of this evaporated moisture some 1430 cubic miles falls annually on the continental United States as rain and snow; about 390 cubic miles gets back to the sea via surface streams, and 40 cubic miles through underground passageways. There is a constant interchange of water between surface and subsurface, but the trend is always seaward and the distances covered are impressive. Much of the river water that reaches any ocean has flowed hundreds of miles and has been an erosive agent all the way.

The Mountain Brook

If any large river in a humid region is followed upstream, it will probably lead scores or even hundreds of miles inland—past many junctions with tributaries, through terrains rugged and smooth—to its origin on a highland. The origin may be a spring or perhaps the melting lower tip of a glacier. The flow of water here usually looks insignificant. Somehow this little stream manages to grow during its travels to a great river before it meets the sea. The explanation for this remarkable fact is inherent in the nature of rivers: they are fed not only by the flow at the origin, by direct rainfall, and by surface drainage, but also—and most of all—from supplies of water underground.

The two million cubic miles of water in the crust is water that fell as rain or snow and has percolated down in tiny zigzagging streams between rock and soil grains, between rock strata, and through rock fractures.

Downward movement of such water continues until the top of the saturated zone is reached; below this level, called the water table, all available spaces are filled. At the water table the water spreads out and, following paths determined by the rock structures, continues to seek lower levels. Here and there some of the water emerges into basins to form lakes or swamps, and into ravines or valleys to feed streams. Thus streams, by cutting gullies, ravines, and valleys, tap flows of this water, and the deeper they cut the more water they are likely to obtain.

Near its origin on a mountain the typical stream occupies a small gully. This usually widens and deepens downward to become a ravine, and as this leads to lower levels, still deepening, it taps increasing supplies of ground water in the form of springs, and its flow steadily increases. By the time it has reached the trunk valley at the foot of the mountain, it has become a full-fledged brook. In the valley it joins a much larger stream, fed by many brooks, and this stream—better called a river—runs down the valley and out onto the lowland.

The brook on the mountainside, having a high gradient, cuts downward busily. It destroys the rock of its bed by solution, hydration, oxidation, and carbonation. When the brook is swollen from heavy rains or melting, the water rams into rock crevices and sucks on the bottom and walls of the channel. In winter, ice scraping along in the channel adds to the wear and tear. Bits of silt and sand carried in suspension, gravel bounced along the bottom, and cobbles and perhaps boulders rolled intermittently downgrade —these keep up a battering attack on the bedrock of the channel, especially the bottom. Of much importance in this downcutting process is the drilling of potholes by whirling sediments in eddies.

Downcutting by the brook results not only in the deepening of the ravine but also in its lengthening upslope. As the ravine deepens, the gully that is its upper end is extended upward. Above the origin there is no spring, but with each rain or melting the gully is eroded a little deeper, widened, and lengthened headward. As it lengthens it intersects smaller gullies and captures their flow. As it deepens, it may tap additional springs, each of which in turn becomes the origin. Upward lengthening may thus continue until the gully reaches the top of a ridge. There, with the upper end of a gully on the other side of the ridge, it may form a col, or saddle.

Since most of the brook's cutting is downward, one might expect the

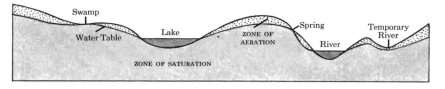

A water table: The diagram is simplified; a water table is usually more irregular. Impervious rock strata may form zones of saturation within the zone of aeration.

Potholes in a riverbed: In Henrico County, Virginia, pothole-drilling has been important in the James River's downcutting. Rock fragments in potholes (as at *lower right*) serve as abrasives.

result to be a cleft as deep as it is wide. Actually the cross section of a ravine, especially after it has become well developed, is likely to be that of an open V. The reason for this form is that as the ravine is cut deeper, its walls are worn back by weathering, mass wasting, and slope wash. The V tends to be widest at the top because it is there that the attack on the walls has been going on longest.

The channel is likely to be choked with large-caliber debris, some of it derived from the brook's own work and some from the attack on the ravine walls. The rapid rate of downcutting relative to sideward erosion keeps the walls steep; and so, as rock fragments on the walls become loosened, they readily roll, slide, or creep down to the streambed. This debris represents a heavy supply of tools for the abrading and hammering work of the brook. The more rapid the downcutting, the faster this debris is produced; and the faster the stream flows, the more effectively it can employ these tools—not by a simple arithmetic progression but geometrically. The ultimate control of the stream's rate of downcutting is gradient.

The swifter the brook, the abler it is to cut the ravine straight downslope. Rock resistance does, however, tend to influence the cutting. Streams

A young river: This stream in the Blue Ridge Mountains of Virginia has not been following its present course long enough to eliminate the irregularities in its bed. In the foreground is Linville Falls.

tend to follow the zones of weaker rock, and if these zigzag, so do the streams. Where the water flows over a rock stratum that dips toward one side of the streambed, the stream tends to shift toward that side, undercutting and oversteepening the ravine wall.

Outcrops of relatively strong rock in the streambed may produce rapids or small waterfalls. If the strata are turned upward, the stream may be deflected by them, running around the stronger, higher-standing ones and cutting down into the weaker ones. As, in the course of time, the highland is reduced by erosion and new patterns of rocks are exposed, the patterns of drainage become adjusted accordingly.

Such, essentially, are the mechanics of stream work on humid highlands in temperate and tropical regions. On arid or very cold lands, such as those of the Basin and Range Province in the western United States and the Canadian Rockies, there are differences due to the low rate of

A stream following a dike: The Flume, in Franconia Notch, New Hampshire, has cut down through a relatively weak basaltic dike in strong granitic rock. Rectangular blocks of dike rock are visible on the riverbed.

precipitation or the freezing of most of the available moisture. Streams on desert highlands, though usually only temporary, have tremendous destructive power because of the lack of vegetation cover. The alluvial fans and cones often seen at the foot of desert ravines demonstrate the transporting abilities of these streams. But on cold highlands most of the moisture remains in the form of ice, and stream work is minimal. Here frost-wedging may account for most of the rock destruction. At the foot of the ravines one often sees deep cones of weathered debris, some brought down by meltwaters and much by mass wasting.

The Valley Stream

Where a mountain brook joins a valley stream, the scene changes abruptly. The valley is much deeper and wider than the ravine; it is in fact

Cut by a young, swift stream: Ausable Chasm, near Keeseville, New York, has been cut by the Ausable River since the Pleistocene glacier melted away about 15,000 years ago. Rectangular jointing of the sandstone has favored the formation of nearly vertical walls.

a ravine in which a stream was working perhaps a hundred thousand years before the brook was born, and thus it represents a much longer erosion span. The gradient of the valley floor is much gentler than that of the brook bed, yet the flow may still have a high velocity, perhaps five to ten miles per hour. The river contains a very large volume of water, thanks to its multitude of tributaries and to its downcutting, which has reached the saturated zone along most of its course.

The valley stream flows more smoothly than the brook because its channel is smoother. During its longer working span the river has been grading its bed and cutting off projections from the sides of its channel. Humps and projections have been eroded faster than the rest of the trough. The result of grading is the elimination of rapids and waterfalls, and also some straightening of the channel.

But this valley, almost as much as the ravine, may express varying rock resistance. Here and there the river curves to take advantage of a fault zone or a dike of weak rock. At some points it may cascade over a step or bench representing the edge of a resistant stratum. Along some stretches the valley walls may become very steep—an indication that unusually strong rock resistance has retarded valley-wall erosion relative to downcutting. Elsewhere the valley may open rather widely: here its walls are of weaker rock. But such irregularities diminish noticeably as the valley lengthens toward the lowland. Quite clearly, the river with time makes its course more regular.

Waterfalls in Yosemite Valley, California: A Pleistocene glacier of tremendous depth cut the valley much deeper than thinner glaciers cut the tributary valleys. Hence streams issuing from the tributary valleys today have a long fall to the canyon's floor.

One notices that this large stream has more tributaries than the brook. Being older, it has had more time to cut headward and intercept streams at higher levels. Drainage to the river is better organized, more efficient, than to the brook. If the landscape could be seen from an airplane, the drainage into the brook would seem almost random, with much zigzagging and parallel running by the gullies, but the drainage into the river would appear to be better integrated, with the brooks taking more nearly direct courses toward the river. It might be noticed also that whereas tributaries often enter the brook as waterfalls, the brooks usually enter the river at its own level—evidence of a long period of mutual accommodation.

At the junction of the brook with the river there is much large-caliber debris—cobbles and boulders which the brook with the aid of gravity could move downslope, but which the larger stream with its slower velocity and diminished aid from gravity has been slow to clear away. Despite appearances this rock waste is being reduced by solution and abrasion and is being moved downstream at times of high water. On the average, the caliber of sediments in the river is smaller than in the brook. The river has had more time to reduce this material; also, because the valley walls are not so steep as the ravine walls, debris moves down to the riverbed more slowly and suffers more attrition on the way down. Where the valley is wide, the proportion of large-caliber debris in the channel is likely to be low.

Whereas bedrock was exposed along most of the brook's bed, it is rarely visible in the river's channel. Here it is mostly covered with sand, gravel, and cobbles. Under ordinary conditions this material protects the bedrock. Only during flooding can it be picked up and used destructively.

The manner in which sediments are distributed on the riverbed is significant. They tend to accumulate in depressions where the stream loses velocity and, therefore, carrying-power. By contrast, little sediment is seen on humps in the channel, because to get over these humps the water must flow faster, just as the air flows faster over the cambered upper surface of an airplane's wing.

Sediment accumulates also where the stream widens, because as the waters spread they again lose velocity. Where the river curves, debris collects on the inside of the bend because the water there is slower than that farther out. At every point on the watercourse, both the amount of sediment that accumulates and the calibers represented tend to be determined by the velocity of flow.

Along stretches where the trough widens considerably, the loss of velocity may allow vast amounts of rock waste to accumulate. The resistance presented by this mass may cause the river to divide into a number of smaller streams which, by virtue of their narrower courses, flow faster and thus succeed in forcing a way through. The result of this process, which is called braiding, is the formation of elongated islands in the river. Braiding

Alluvial cones: In Banff National Park, Alberta, rock waste from intense frost action is washed down through ravines to form cones at the foot of the peaks.

is likely to be seen where rock waste has accumulated faster than would be usual for the rate of downcutting. This may happen in valleys of arid regions because of heavy sheetflooding, and on cold terrains because of the high rate of rock-waste production—such as by frost work and glaciation—relative to the carrying powers of the streams. Braiding may occur also, as we shall see, on any lowland where sediment deposits are deep and the gradient is slight. Braiding is spectacular on the flood plains of the North Platte in western Nebraska and on the Yukon Flats in Alaska.

Back on the mountainside it was noticed that the brook's channel fully occupied the bottom of the ravine. But the river's channel at many points may not fully occupy the bottom of the valley. Often there are flat, terrace-like deposits of sediments flanking the channel—deposits built up by flooding. The farther one travels downstream, the wider some valleys become and the more extensive are the deposits. Eventually, toward the foot of the valley, they are extensive enough to form a broad, nearly level valley

Junction of two major streams: At Harpers Ferry, West Virginia, the Shenandoah (*left*) and the Potomac meet at about the same level—evidence of a long period of mutual accommodation.

floor, only a small part of which is occupied by the channel. The channel tends to swing from one side of the valley to the other, as if unable to maintain a straight course through the increasing masses of waste. Here, it appears, most of the sediments are being moved very slowly. Mass wasting and slopewash are bringing down material from the valley walls rapidly relative to the carrying powers of the river. The bedrock is deeply buried. Downcutting has virtually ceased.

The Lowland River

The river eventually issues from the highland valley onto a broad lowland. This may be visualized as a wide, nearly flat terrain such as those of the Great Plains or the Gulf Coast of the United States. Here the valley walls may be miles from the channel. Swelled by waters from many tributaries, the stream now ranks as a major river. The sediments of ages, scores or hundreds of feet deep, are waterlogged, and the river represents not an independent stream fed by springs but an exposure of a vast reservoir of ground water barely beneath the surface. Thanks to this water supply, the river may have a fair velocity, yet it seems grossly inadequate to the task of keeping the sediments moving. It meanders across the lowland, heading one way and then another, doubling back on itself and then turning forward again, mile after mile. Only the minimum of erosion is occurring here under normal conditions. But appearances are somewhat deceiving. During one flood the transport of sediments by waters overflowing the channel may far exceed the annual accomplishment of the river back in the highlands.

Meanders are common, but not well understood. Tests with laboratory models indicate that any stream will weave when flowing through a mass of sediments. The weaving starts after turbulence has created unevenness in the stream bed. In nature even fairly rapid streams moving through channels in rock have a tendency to meander, sometimes without apparent influence by variations in rock resistance.

Whatever their origin, meanders on a flood plain rank as the most changeable of rivers. The tendency to undercut the outside bank downstream from the beginning of a curve causes downstream migration of the curve. Under flood conditions an entire loop may be cut off by waters rushing directly across the divide.

Parts of loops, called oxbows, often become sealed by sediments, and

A braided stream: The upper Firth River in Alaska is much braided in its course through abundant glacial deposits in the valley.

if their bottoms are below the water table they become oxbow lakes—lakes that may last only until the next big flood. Seen from the air, many flood plains display a multitude of oxbows, most of them filled with vegetation and now scarcely recognizable. One wonders how many property owners in the old days took the stream for a boundary line.

Another feature of the flood plain is the natural levee—a parapetlike ridge of sand and mud dropped as flood waters overflowing the channel spread out and abruptly lost speed. Natural levees like the artificial variety form bulwarks against further floods that are not high enough to sweep over them. They are, however, only a temporary safeguard, because if flood waters are confined within the channel they dump their sediments there, raising the level of the river; then the levees too have to be raised, and after a time the river is flowing well above the surrounding countryside—obviously a grave hazard in flood time. The only solution then appears to be flood control upstream.

Although on a flood plain downcutting has virtually ceased, degradation is continuing. As long as the river has the slightest gradient, it keeps carrying rock waste seaward. This waste continues to be brought from upstream and comes also from the erosion of the nearest highlands, even if

Loops of the White River in Arkansas: This wide flood plain near Des Arc, with its meanders and oxbow lakes, typifies a stream in old age.

An arcuate delta in Cook Inlet, Alaska: Innumerable small distributaries have created here a delta with only minor indentations on its front.

these are many miles away. The slow downward creep of sediments continues, at least in principle, as long as one grain remains above sea level.

What happens, finally, to sediments that are delivered to the ocean? The answer depends primarily upon the nature of the coast. If the seaward end of the river valley has been drowned—as many have been—by a rise in sea level or by a subsidence of the coast itself, the river-borne sediments are dropped on the valley bottom as the gradient with respect to sea level nears zero. This is the situation along drowned highland coasts such as that of Maine. Where the water offshore is shallow and the bottom is gently sloping, as is true off the East Coast southward from New Jersey, the sediments are distributed along the shore by waves and currents to form beaches and bars. Finally, where rivers empty into shallows where the water is relatively quiet, as it is along the Gulf Coast, the material accumulates to form deltas.

On reaching its delta a river divides into separate streams which, following channels cut during floods, carry rock waste to the delta's edges, dumping it there to form inclined layers called foreset beds. Mingled with these sediments are precipitates from chemical reactions between the river water and the sea water. With the new materials brought by each flood, the delta grows outward. If it grows substantial enough, the sea bottom under it will sink under the weight.

Deltas were so named by the Greeks because of their approximate similarity to the Greek letter Δ. Actually most deltas, such as those of the Nile and the Rhine, have a rounded outer edge, not a straight one. These deltas, generally termed arcuate, result from close spacing of distributaries, so that the deposits form a nearly unbroken front.

Arcuate deltas have been said to form where the sediments are coarse and allow relatively free movement of the water, so that a multitude of distributaries can develop. A less common kind of structure, the bird's-foot type, is represented by the Mississippi Delta. It consists of smaller-caliber, more densely packed sediments which force flood waters to concentrate into a few strong, widely spaced streams. A delta with long promontories results.

Deltas of large rivers are built rapidly. The Mississippi and Po deltas are now extending themselves at the rate of about two hundred feet per year. The delta of the Po has grown about fourteen miles out into the Mediterranean during the past eighteen hundred years, making the ancient seaport Adria an inland city. The Nile Delta is building outward at the rate of about twelve feet per year.

Inland deltas are small but numerous. These take shape where heavily loaded streams empty into lakes or into other streams. Many were formed by river-borne glacial debris during the Pleistocene meltings: for example, the deltas in the Hudson River near Bear Mountain in New York. Two of the largest lake deltas are those built up in Lake Mead, Arizona, by the Colorado River and in Lake Geneva, Switzerland, by the Rhône.

Guidance by Structure

Especially in highlands, river work is guided by rock structures—not only by local features, such as dikes and outcrops of varying resistance, but by regional structural patterns. Streams can be classified according to the courses they have developed on broad structures.

Some streams originally develop courses according to the dip, or inclination, of rock strata beneath them. These streams, called consequent, may be seen running down recently formed slopes, such as coastal plains or the sides of relatively young mountains like the Sierra Nevada of California and the Wasatch Range of Utah.

Certain streams find courses along belts of weak rock exposed by the truncation of the limbs of folds. These streams are called subsequent. Since they run parallel to the axes of folds, the valleys they cut are often called longitudinal, or "strike," valleys.

Valleys that are cut along the axes of truncated anticlines—that is, upfolds—are called anticlinal, and those following the axes of synclines, or downfolds, are synclinal. (Folds are illustrated on page 107.) These valley types are common in the Folded Appalachians.

Obsequent streams have a direction opposed to the dip, or inclination, of strata. This pattern is seen in the Catskill Mountains of New York where Kaaterskill and Plattekill creeks run down the eastern scarps of westward-dipping strata. Obsequent streams are familiar on the sides of basins formed by the truncation and hollowing out of a dome, such as the Nashville Basin in eastern Tennessee and the Weald in southern England.

The resequent stream is likely to be identifiable only by a field geologist, but must be mentioned here to complete the roster. This stream, like the consequent type, follows the dip of a stratum, but differs in that it originates on this stratum only after erosion has removed overlying rock. Such streams are characteristic of slopes resulting from the erosion of tilted strata.

Streams commonly develop on slopes created by faulting. Those that flow in the direction of the original dip are called resequent, and those that flow the other way are obsequent.

Some streams occupy natural troughs or trenches that have been formed directly by faulting and are known as rift valleys or grabens (from German, "grave"). One of the most famous of all stream-occupied grabens is the Rhine Valley in Europe. Another such feature is Death Valley in California—a sunken land so arid that its rivers, the Amargosa and Furnace Creek, die in the sands.

Cross profiles, too, show structural effects. Many steep-sided valleys, especially those cut into sedimentary strata, display in cross section "ribbed" walls, the ribs being due to varying resistance, and their arrangement being an aspect of structure. The lower portion of a valley cross section may show structural benches, or rock steps, of similar origin.

A valley that tends to run across strata at an angle to the dip usually has a steeper wall on the down-dip side, because of undercutting. Again, if rocks on one side in any valley are stronger than those of the other side, the wall with the stronger rock is likely to be the steeper one.

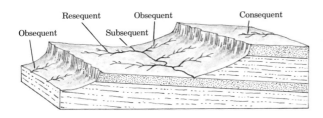

Modes of origin of streams: Watercourses develop wherever there is a water supply to an inclined surface. The direction taken tends to be the direction of steepest gradient.

When a valley has been cut between strata that are nearly vertical, the walls ordinarily are very steep, even if the rock is weak. Streams easily penetrate the strata, and thus downcutting dominates. Subsequent streams in belts of weak rock along the edges of the Folded Appalachians offer good examples.

Rock structure accounts for many waterfalls. One of the most dramatic specimens is Kaietur Falls on the Potaro River in Venezuela, which leaps 740 feet over the edge of a stratum of strong conglomerate overlying weaker sandstones and shales. The two falls of the Yellowstone in Yellowstone National Park occur at the edges of strong rhyolite lava flows that overlie deposits of weak volcanic debris.

Niagara Falls, near Buffalo, New York, is located where the river, flowing north over a strong limestone, reaches the end of this formation and continues on weak shale. The accompanying diagram shows how the cliffs over which the waters plunge are kept steep by undermining; turbulent waters keep digging out the shale from beneath the limestone. As overhangs of limestone break off, the falls gradually moves upstream, the present rate being a little under four feet per year. When, a few thousand years hence, the river cuts through the bottom of the limestone stratum and reaches the shale beneath it, downcutting will accelerate, no more overhangs will form, the falls will become a mere rapid, and the Mecca of honeymooners will be no more. By that time, perhaps, a more impressive waterfall will have been constructed by a concessionnaire.

Waterfalls and rapids resulting from a very large structural feature are seen along the Fall Line in the eastern United States. This line is the approximate boundary between the Atlantic Coastal Plain and the higher lands of the adjacent interior, called the Piedmont Upland. The boundary runs along the foot of the slopes formed by the Upland's resistant rocks; eastward lie the loose, younger sediments of the Coastal Plain.

The Colorado River in its Grand Canyon also shows structural controls. At the upstream end, the river is cutting only through sedimentary rocks and has roughly the profile of a very wide V, with terraced sides. Farther downstream it has cut through the lowest layers of the sedimentaries and

The rock structure beneath Niagara Falls: The existence of the falls depends upon the superior resistance of the overlying rock layers relative to those beneath.

has begun slicing into the more resistant granites and schists beneath them. At the level of the strong rock the canyon narrows sharply. The nearly vertical, armorlike walls, towering above the slim torrent, contrast utterly with the upstream panorama of a slower river in a wider channel, flanked by deeply gullied, crumbling, rapidly retreating valley walls.

The contrast will not last forever. Downcutting will wane as gradient is reduced and as backwasting of the canyon walls continues. Debris will accumulate in the channel at a faster rate relative to the ability of the river to carry it downslope, and the canyon will start filling. Over the next million years or so, unless a new uplift renews the downcutting, the dramatic expression of changing rock strength will disappear. A much wider, much shallower Grand Canyon, its towers and pinnacles gone, the river winding slowly through masses of stalled erosion debris—this will be the sight viewed by the tourists, whoever or whatever they may be, of that remote day.

Rivers That Flowed Uphill?

On occasion one sees a river valley that has been cut directly through a mountain range transversely, in apparent disregard of rock resistance and structure—and of gravity. Some such valleys have been cut by headward erosion along a zone of weakness represented by a tear fault or a dike. More common are those cut by superposed or antecedent streams.

The superposed (or superimposed) stream appears to have flowed up one side of the mountain range and down the other, maintaining this impossible course until the range was cut through. Contrary to appearances, however, gravity has not been defied. To understand what has happened we must imagine the river flowing across the terrain as it was long ago. Erosion had reduced the region almost to a plain. Beneath the plain were the stumps of an ancient mountain range, buried under erosion debris. As renewed erosion reduced the terrain, it exposed the old mountain mass. Since the river's course lay directly over this mass, the river cut down through the rock like a bandsaw, while other streams shaped a new

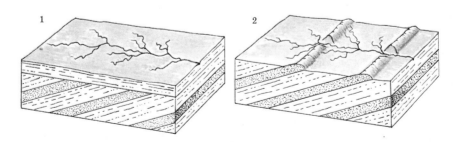

Superposition of a stream: Where a stream has established a course, it may be able, as here, to hold that course even after differential erosion has altered the conditions that determined the course originally.

topography according to the structural controls. The mountain rock was more resistant than the sediments that had buried it, but the river maintained its course by narrowing its channel and speeding up.

The United States, particularly the Eastern highlands, is rich in superposed rivers. A relief map will show such rivers cutting through the Appalachians—for example, the Delaware, the Susquehanna, and the Potomac. The passages are popularly known as water gaps. Streams cutting across the Blue Ridge in Virginia are superposed, and so, possibly, is the Hudson where it breaches the Hudson Highlands. The Connecticut River is superposed on the Holyoke Range in Massachusetts, the Snake River on the Teton Range in western Wyoming, and the Shoshone River on the Absaroka Range in the same state. Famous superposed streams in Europe include the Meuse where it crosses folds of the Ardennes. In England superposed drainage can be observed in the Lake District.

The antecedent stream likewise cuts transversely through highlands, but its history is different. During the recent geologic past a portion of the lowland over which the river was flowing gradually began to rise. The uplift was slow enough so that the river, instead of being diverted, was able to maintain its course.

Since antecedent streams can be identified only where uplift has been relatively recent, examples are limited. In the United States one river that may be antecedent is the Columbia in its gorge through the Cascade Mountains in Oregon. Another is the Santa Ana where it crosses the northern Santa Ana Mountains in southern California. A possible antecedent stream in Europe is the Danube where it cuts through the Alps of Transylvania. Some geologists also include among antecedent streams the great Himalayan rivers—the Indus, the Brahmaputra, and others. The rapid cutting that has been necessary to keep pace with the uplift has produced

A superposed river: After its junction with the Shenandoah River (at *right* in photograph) at Harpers Ferry, West Virginia, the Potomac pours through its water gap in the Blue Ridge, seeking the distant Atlantic.

An antecedent stream? The Columbia River is believed by some geologists to have held its course through the Cascade Mountains while they were being built up. The view is eastward from Crown Point.

very narrow gorges of awe-inspiring depth. In Kashmir the gorge of the Indus is nearly twenty thousand feet deep.

Drainage Systems

Every stream is a member of a drainage system in which many are linked, and in which any change that affects one may ultimately affect all the others too. Even a local change, such as a fault that creates a waterfall or a landslide that dams a valley, starts reverberations that may continue almost indefinitely both upstream and downstream, eventually involving the whole drainage system. One could almost say that a pebble tossed into any river will change the drainage of the continent.

The sensitivity with which streams can become adjusted to one another was recognized as early as 1802 by John Playfair, the popular interpreter of James Hutton, in his *Illustrations of the Huttonian Theory of the Earth*:

> *Every river appears to consist of a main trunk, fed from a variety of branches, each running in a valley proportional to its size, and all of them together forming a system of valleys, communicating with one another, and having such a nice adjustment of their declivities that none of them join the principal valley on too high or too low a level; a circumstance that would be infinitely improbable if each of these valleys were not the work of the stream which flows in it.*

Playfair was describing a drainage system that had not suffered the upsetting events experienced by most streams of reality, but no one has described more neatly the nature of stream interdependence.

Stream systems, like individual streams, show a logical development. In the early phase—let us say on a newly uplifted surface—the streams are widely spaced with few tributaries: the texture of the system is coarse. Valleys are deep and narrow, with high, wide divides between them. Junctions are rare, seeming to result from chance convergence of zigzagging courses. But with time the streams by headward erosion acquire more tributaries, and along these we find gullies strung in increasing numbers. The texture is now fine. Stream junctions are more numerous; divides have been worn lower and narrower; drainage is integrated. But this phase of maturity gradually passes into decline. As the streams lose gradient and the landscape generally is reduced, headward erosion and downcutting slacken, sediments from valley-wall erosion accumulate, divides are all but leveled, and gully systems are shallow, poorly connected, and low in collecting efficiency. The drainage texture has become coarse again.

Stream systems become integrated by the process called stream capture, or stream piracy. By headward erosion and by swinging, streams cut away the relief that separates them, and wherever two watercourses meet one of them takes all or most of the water. The process can be watched in miniature on any slope of loose soil during rain. On a grand scale it has altered the aspect of entire regions by interrupting the normal development

A drainage system: The relationships of integrated stream courses on a highland are demonstrated in Angeles National Forest near Los Angeles, California.

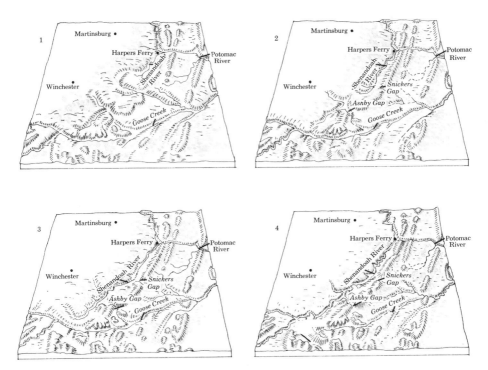

Stream captures by the Shenandoah River: Cutting headward toward the southwest, the river has beheaded, one by one, streams that once cut through the Blue Ridge.

of valleys. Gain or loss of water supply causes, respectively, increased downcutting or increased deposition. Valleys in which rivers have worked for thousands of years may go dry, and in other valleys the streams may gain greatly in potency.

Capture usually is accomplished by means of superior gradient. The pirate stream may, for example, have a course down the steep face of a cuesta. Streams on the backslope, having but gentle gradient, become the pirate's victims. This kind of capture is seen along the eastern edge of the Appalachian Plateau—for example, where Kaaterskill Creek flows down the Catskill escarpment. This stream has at two points captured tributaries of Schoharie Creek, a stream of lesser gradient on the backslope of the Catskill upland.

In upland terrains of limestone, underground streams may do considerable tunneling by solution. Wherever a surface stream cuts through to one of the underground water courses, the latter takes the water supply. Such captures account for the "lost rivers" of limestone regions such as those of Kentucky and southern Indiana. West of Bloomington, Indiana,

A wind gap: New Market Gap in Massanutten Mountain, Virginia, was formerly a water gap. The Shenandoah River captured the water supply of the transverse stream.

underground water courses have captured about fifteen square miles of the surface drainage of Indian Creek and diverted these waters into Clear and Richland creeks.

Piracy is common where transverse streams are being superposed on a buried folded structure. At first the streams hold their courses right across the folds, cutting water gaps. Meanwhile, as the buried structure becomes exposed, subsequent valleys are cut in belts of weak rock parallel to the fold axes. Being in weak rock, the subsequent streams cut faster than the transverse ones as they extend themselves headward. When they meet the transverse streams they have steeper gradient at the points of junction. Accordingly they may take all or most of the water supply, leaving the transverse valleys downstream from the junctions either dry or with much diminished flow. The over-all result is a landscape of long subsequent valleys, few if any transverse streams, and "wind gaps" where the water gaps used to be.

Such has been the story of the Blue Ridge region in Virginia. Perhaps 15 to 20 million years ago, a group of streams flowing seaward from the Appalachians in what is now western Virginia were being superposed on the Blue Ridge. This ancient range, buried in debris from an earlier phase of Appalachian erosion, was being uncovered in a new cycle started by renewed uplift. Transverse streams were cutting water gaps through it. Meanwhile the Shenandoah River, a young tributary of the Potomac, was cutting headward through a belt of southwestward-trending weak rock west of the Blue Ridge and south of what is now Harpers Ferry, West Virginia. The Shenandoah lengthened rapidly because it was in weak rock and had behind it the steep gradient of the fast-cutting Potomac; but the rivers crossing the Blue Ridge were cutting only with difficulty through the strong metamorphic and volcanic rocks. Accordingly, over a period of a few million years, as the head of the Shenandoah reached each of the transverse streams its steeper gradient enabled it to capture the water supply. Thus the portions of the transverse valleys lying between the Shenandoah and the Blue Ridge went dry, and the water gaps through the ridge became wind gaps—notably Snickers, Ashby, Manassas, Brown, and Rockfish. Today the Shenandoah is still cutting headward, and within the next few million years it will, barring interference, behead first the James River and then the Roanoke.

In England comparable acts of piracy have been committed by the river Ouse where it has cut headward in a northwesterly direction along the east side of the Pennine Hills. It has pirated, one by one, three eastward-flowing streams—the Nidd, the Ure, and the Swale—which formerly ran the full course to the North Sea.

Drainage Development

A stream system is influenced strongly by variations in the rock. Homogeneity of the rock and high resistance together favor wide spacing, as in granitic parts of the Hudson Highlands and the New England uplands, because on such rock the running water finds few lines of weakness to exploit. High permeability of rock, as in the sinkhole region of Kentucky or the chalk terrains of England, also tends to favor wide spacing, because the rock absorbs rainwater before much of it can be gathered into gullies and ravines. Weak and less permeable rocks, on the other hand, are associated with closer stream spacing, which in its extreme form yields badland topography such as that of South Dakota and Wyoming.

Maps show that most drainage systems follow certain standard patterns at least approximately. Most familiar is the dendritic pattern. Developed especially on homogeneous rocks, such as granite or basalt, and on broad expanses of horizontal sedimentary strata, it is a random arrangement that shows little structural influence. Sometimes it occurs where a drainage system, originally developed on one rock structure, has persisted even after erosion has exposed a different structure. Dendritic drainage is familiar on the granitic hills of the northeastern United States and on the nearly flat-lying sedimentaries of the interior.

Where mainstreams following the trend of folded structures are fed by tributaries coming down the sides of parallel ridges, the trellis pattern may result. Occasionally one of the mainstreams bends at a right angle to flow through a gap in a ridge and join the mainstream in a neighboring valley. If erosion on the terrain is advanced, so that the number of tributaries is diminishing, the trellis pattern may be giving way to a parallel arrange-

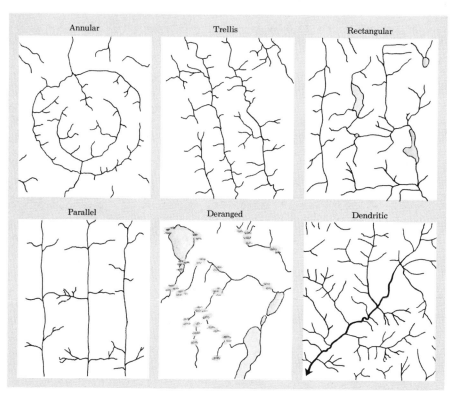

Drainage patterns: Each stream in a drainage system tends to follow the path of steepest gradient. Especially on highlands the system expresses rock structures.

ment. Trellis drainage is well developed along the folds of the Appalachians, most noticeably in central Pennsylvania and southeastern New York.

River courses may follow master joints and faults. Rectangular patterns apparently due to faulting are formed by valleys in the Adirondack Mountains of New York. Lake Placid clearly shows this pattern, and to the east, a valley featuring right-angle turns has been cut by the Ausable River.

On domelike structures drainage may show either radial or annular designs. If the rock is relatively homogeneous, the streams tend to follow the dip radially. This pattern is typical on volcanoes, such as Mount Rainier, and occurs also on alluvial fans and cones. If, on the contrary, exposures of alternating weak and strong strata form rings around the dome, some streams are likely to be found cutting into the weak rings and creating the annular pattern. Such a ring is represented by the Red Valley, which nearly encircles the Black Hills dome, and by streams around the Weald, the truncated dome in southern England.

Annular drainage may result also from ringlike exposures of weak rock on slopes of structural basins. But if the rock is homogeneous, a centripetal arrangement is likely; that is, the streams will flow down the sides toward the center of the depression. This pattern is seen in the Paris Basin in France, where streams run generally toward Paris from the north, east, and south, eventually emptying into the Seine.

Parallel drainage develops where steep mountain slopes on homogeneous rock give out onto broad plains. This pattern occurs among large consequent streams flowing eastward from the Rockies onto the High Plains, and those that flow down the western slopes of the Sierra Nevada. The same pattern appears on the Atlantic and Gulf coastal plains, where streams flowing seaward through unconsolidated sediments meet little variation in resistance and thus maintain the relatively direct route favored by gravity.

Some other drainage patterns represent combinations of the various standard forms. Still others are standard forms that have been altered by some kind of interference, such as uplift, glaciation, or volcanism.

Rivers Young and Old

To follow a river through space is like following it through time, from youth to old age. Gradient is the elixir of youth. Between the highlands and base level, gradient wanes from the near vertical to almost nothing, and the working powers of the stream wane accordingly. Originating in youthful, erratic haste, the river swells and deepens, passes through a mature stage of more deliberate flow through a widened scene, and approaches its final destination, the sea, as an old stream with powers no longer adequate to its task. In the highlands the river's valley is youthfully rugged and steep, in the midlands wider and better graded, but near the sea its contours soften toward no relief at all.

The metaphor has a double sense. Not only does the river age between its origin and its mouth, but if one could watch any of its segments in the highlands or midlands for a few hundred thousand years, one would see this segment itself aging.

Old age for both river and valley begins at the sea and, barring the unusual, creeps methodically upward toward the highlands, and when it has arrived there the highlands are gone.

As any stream is followed toward base level, its "age" is likely to keep changing to older or younger and back again as new conditions—especially those of rock structure—are encountered.

The Connecticut River, for example, rises in the Connecticut Lakes in northern New Hampshire, then hurries south through a narrow valley in strong rocks as a youthful stream. Reaching the zone of weak red sandstones and shales in western Massachusetts, it flows more leisurely through a much-widened, mature-to-old valley, then resumes a faster flow through the resistant basaltic ridge known as the Holyoke Range. The river widens again in the relatively weak sedimentaries of northern Connecticut, then narrows once more as it breaches a wide barrier of strong crystalline rocks before reaching Long Island Sound.

The Mississippi River follows a lowland virtually all the way from its origin in Lake Tasca, Minnesota, to the Gulf of Mexico. Its moderate gradient down to the vicinity of Cairo, Illinois, makes it generally a mature stream that far. But between Cairo and the Gulf, a distance of some six hundred miles, the river has a fall of only about three hundred feet, or an average of six inches to the mile. Accordingly, this part of the Mississippi Valley has become a showplace of fluvial old age. The valley, fifty to seventy miles wide, is essentially a flood plain. All manner of meanders and cutoffs appear, and the resulting oxbow lakes are multitudinous. Here also are classic examples of Yazoo-type streams: tributaries that follow the main river on a parallel course for long distances before they finally succeed in penetrating natural levees to make the junction. The Mississippi south of

A mature stream: The Hudson River in its passage through the Hudson Highlands is a stream at maturity. The view is northward from Bear Mountain, New York.

A young stream in shale at Watkins Glen, New York: Joint patterns and variations in resistance of the rock account for the irregularities in walls and streambed. An underground stream emerges from the cliff at left. Note the potholes in the streambed.

Cairo is not presently downcutting (except as part of the slow general lowering of the entire landscape), but the depth of sediments in the valley —averaging over two hundred feet along some stretches—and the Mississippi Delta testify to the stupendous work of transportation done by the river in the past, especially during the late-Pleistocene meltings.

Among the principal rivers of the world, the Yukon is of special interest because of its two-thousand-mile length and its rather orderly progress through the classic fluvial stages. Its headstreams originate in the Chilkoot Pass neighborhood of British Columbia at an elevation of about six thousand feet. Flowing vigorously northward, the Yukon enters a mature phase,

Fantasy of old age: The Yukon on its famous flats in Alaska seems to have explored all the possibilities for a meandering stream. A high-altitude view of this region appears on page 261.

with developing flood plains, near Dawson in the Yukon Territory, and then goes on to an old age in the vast Yukon Flats, where its flood plain becomes ten to twenty miles wide, and the river braids its way through the sediments with wide wandering. West of the flats the waters speed up and the valley narrows as several zones of resistant rock are traversed. Then, below its junction with the Tanana River, the Yukon meanders through a vast meadowland, virtually at base level, before reaching its delta and the Bering Sea.

Interruptions of Development

All rivers are subject to interruptions of their normal development, and rare is the one that evolves through all the classic phases without substantial upset. Any event that directly alters the gradient of a river or its water supply, directly or indirectly, may have important results.

Many streams have become blocked by a landslide, a lava flow, a delta or fan built by a tributary, or a barrier beach on a seashore. Blockage diverts the flow and may cause ponding. Upstream the gradient is reduced, and deposition accelerates. (This makes a problem especially in the West where engineers dam streams that carry very large amounts of sediments.) Just below a point of blockage, however, flow accelerates and rapidly cuts back into the obstruction.

Landslide blockages are common in mountains. Among notable slides of the past was the 1927 slide in the Gros Ventre Mountains, in Wyoming, which blocked the Gros Ventre River to form Slide Lake. In 1959 a heavy slide in West Yellowstone ponded the Madison River to form Earthquake Lake.

Blockage by a lava flow in Lassen Volcanic National Park is responsible for Snag Lake. Blockages of several small streams by a barrier beach can be seen near Oceanside, California.

Wherever the crust is unstable, normal stream activities may be disturbed by faulting. In California, where horizontal displacements have been occurring along the San Andreas Fault, stream channels crossing the fault have been cut and offset. Faulting associated with volcanic activity, as in Hawaii, Iceland, and East Africa, frequently deranges streams and makes waterfalls. Africa's Victoria Falls is formed where the Zambesi River plunges off a basalt plateau into a graben.

In both northern and southern hemispheres, the advance of the Pleistocene ice sheets gouged out valleys and scoured the plains. As the ice melted, it left an enormous litter of rock waste which was scattered by meltwaters. Many streams were deranged by the advance of the grinding ice or half-choked by the debris dropped later in their channels. They are still readjusting today.

Changes in climate usually mean a gain or loss of water supply for rivers. Reduction of water supply accounts for what are called misfit streams—streams in valleys that show great sediment accumulations and other evidence of heavy flow in the past. One such stream is the Meuse in its course west of Toul Gap in France. Many streams have been made misfits by piracy, and others by the dry trend in climate and temperate regions since the Pleistocene.

The post-Pleistocene rise in sea level has made estuaries of many river mouths. The result has been accelerated deposition far inland. The eastern coast of the United States exhibits many such sea-flooded valleys from the coast of Maine southward. In the Chesapeake Bay area sea waters have penetrated scores of miles inland.

Where a large crustal section is rising, stream development is altered on a grand scale. Visualize, for example, a river winding slowly along a wide, gently sloping valley floor. It is near base level and its work is mainly deposition. But now the crustal block containing the valley is uplifted. At the downstream end a waterfall or rapid is produced, and headward erosion proceeds from this point, deepening the valley. Thus the river is, as geologists say, rejuvenated.

In North America the best known of rejuvenated streams is the San Juan, on the Colorado Plateau. Tens of millions of years ago, while erosion was reducing the plateau, the San Juan established its present meandering

A sea-drowned coast: The mouths of river valleys on the coast of Maine were flooded by the rise of sea level after the Pleistocene. The view is over Bar Harbor.

course. With renewed uplift, beginning two or three million years ago, the river became rejuvenated and its valley was deepened rapidly to form today's canyon.

A number of other streams on the Colorado Plateau have had a similar history. Here and there among the canyons, rejuvenated meanders have cut through the divides to form tunnel cutoffs, and as erosion has reduced the terrain, parts of the tunnels have become natural bridges.

It was the recent uplift of the Colorado Plateau that accelerated downcutting by the Colorado River and created Grand Canyon. Here, near the edge of the plateau, the gradient of the Colorado is relatively steep—as much as seven and one-half feet to the mile—and along a stretch of two hundred miles the river drops from an elevation of about 2640 feet to 1000.

Valleys of rejuvenated rivers often show strath terraces, which are benchlike horizontal platforms cut into the valley sides. Opposite strath terraces represent the level to which the river had cut into the bedrock at the time of rejuvenation. Downcutting accelerated relative to sidewall erosion, and the valley narrowed accordingly, leaving the terraces.

Strath terraces are common along streams in regions of recent uplift, such as the Colorado Plateaus. They are seen also on some lands of older uplift, such as the valleys of Vermont and the lower Thames region of England, where erosion has not been rapid enough to obliterate them.

Some rejuvenated valleys exhibit pairs of strath terraces at several successive levels. These indicate that uplift was intermittent. Intermittency may be indicated also by the presence of "nickpoints," or steplike ledges, in a valley's profile. A nickpoint represents the limit of the river's headward cutting during one distinct erosion cycle.

Strath terraces superficially resemble terraces formed of sediments from flooding; but the strath forms are cut into the rock, not into loose material. These terraces may also resemble structural benches, which are steplike projections from the valley walls due to variations in rock resistance. Such benches are seen above the inner gorge in Grand Canyon. Nickpoints, on the other hand, which are characteristically the sites of waterfalls and rapids, are easily confused with abrupt profile changes due to structural influence. The identification of such features must be left to the expert.

A rejuvenated stream: The Genesee River in north-central New York State was given increased gradient by a recent uplift of this Appalachian Plateau region. The meandering stream then began cutting steep-walled canyons like these near Portage.

A misfit stream: Unaweep Canyon, in Uncomphagre National Forest, Colorado, was originally cut by a powerful river. Much of the water supply was later diverted by the uplift of the Uncomphagre Plateau. The river today looks insignificant.

While the uplift of a crustal block is rejuvenating one segment of a stream, it can be making another segment old before its time. Upstream from a raised block there is a loss of gradient. There the flow slackens, sediments accumulate, and the valley begins to assume the traits of older age. If, on the other hand, a block sinks rather than rises, the river is rejuvenated upstream from the block and loses gradient downstream.

A river disturbed by such changes will by the normal grading process tend to restore the normal state of things. All streams, actually, are doing this in some degree, especially where man has built his dams. In some instances gradient may change too fast for the stream to adjust by grading;

Following a fault: The stream that forms Ribbon Falls in Grand Canyon of the Colorado has cut down fast through a fault zone.

then the waters accumulate against the rise until they can flow over it; or, lacking a basin in which to accumulate, they flow around the rise and establish a new course.

Remarkable examples of combined diastrophic and climatic influences on stream work are seen in the Basin and Range Province. Faulting here within the past 25 million years tilted numerous fault blocks to form highlands and also produced many basins. Water courses lead into these basins, but because of the dry climatic trend of the past ten thousand years, the streams are temporary and lack enough water to fill and overflow the basins. The lowlands therefore are covered deeply with erosion debris which the streams have been unable to clear away.

Peneplains?

The evident fact that landscapes have been degraded miles below their original elevations and later uplifted again suggests the possibility that during past epochs broad areas of the continents have been reduced virtually to base level. Late in the nineteenth century such broad reduced areas, called peneplains (Latin *pene*, almost), were believed to be numerous. Uplifts were generally understood as occurring in distinct episodes between which there was ample time for nearly complete land leveling. Some geologists saw the history of the continents as consisting of clear-cut cycles of uplift and degradation, repeated again and again.

The peneplain concept dominated geology well into the twentieth century, and today many geologists continue to defend it. For peneplains of our time they point to various broad lowlands of slight relief that apparently owe their flatness to erosion rather than to blanketing with sediments or to horizontal rock structures. Cited as examples are parts of the lower Mississippi Valley, the Piedmont of Virginia, and the great meadow of the Yukon. Former peneplains are said to be represented by certain groups of highlands that show tops of about the same elevation. These accordant summits, it is explained, are parts of the surfaces of old peneplains which after degradation were uplifted and then, with their streams rejuvenated, were subjected to a new erosion cycle. On some of these old erosion surfaces leveling had apparently occurred regardless of varying rock resistance; that is, weak and strong rocks had been reduced alike, as would be expected on any landscape eroded down to base level.

A favorite example of an uplifted and dissected erosion surface is the Schooley Peneplain. This is represented by accordant summits scattered from New England to North Carolina and westward at least as far as West Virginia. It is believed to have been raised as much as two thousand feet some 20 to 25 million years ago. Since then the peneplain has been dissected and reduced so thoroughly that today only scattered groups of hills

A peneplain: The Virginia Piedmont, east of the Blue Ridge, is considered to be part of a peneplain created by streams during recent geologic time. A few scattered hills, called monadnocks, survive on this erosion surface.

with accordant summits remain to suggest the former erosion surface. A similar account has been given of many other groups of accordant summits, such as those of the Unaka Mountains of North Carolina, the Central Plateau of France, and the Harz Mountains of Germany.

Today geologists disagree as to the reality of peneplains. All believe that the reduction of vast, lofty highlands by erosion demands time spans of the order of 25 million years, during which there is virtual stillstand—that is, lack of uplift. Some insist that crustal activity through the ages has been sufficient to make any such lengths of stillstand impossible, and that groups of accordant summits actually represent different erosion surfaces. Perhaps the majority of geologists believe, however, that long periods of stillstand have been the rule rather than the exception during Earth's history, and that peneplains have been created again and again.

V

Up, Down, and Sidewise

Hogbacks in the Garden of the Gods: These knife-like ridges near Colorado Springs, Colorado, are the upper edges of limbs of truncated folds of relatively resistant sandstone.

Persons who have lived through earthquakes in regions such as western South America, the Middle East, and Alaska can easily believe that Earth's crust is but a semirigid shell in a continuous state of breakage and distortion. For people in quieter zones, such as the United States east of the Rockies, the crust is still terra firma. Actually, in earthquake zones and quiet areas alike, the crust is always uneasy. Blocks are moving up and down and sidewise, and the bedrock is being folded, fractured, and dislocated. The processes are so slow as to be hardly noticeable except in the earthquake zones, yet in the length of geologic time they produce towering mountains, immense domes, and profound troughs and basins, not to mention multitudes of smaller features.

Wherever masses of bedrock are exposed they are seen to be jointed and often displaced. Most sedimentary rocks, though formed usually in horizontal strata, for the most part, show some folding, warping, and tilting. Among mountains, and especially in the desert where vegetation does not hide rock structures, folds of gigantic size can be recognized, as well as long slopes formed by inclined strata, cliffs made by the tilting of blocks, and basins created by crustal sinking or downwarping. Elsewhere rock distortions are glimpsed in road cuts and quarries, along coastlines, in river gorges, and in simple excavations.

Most landscapes, especially highlands, express crustal distortions in more or less detail. Even where flat-lying sedimentary rocks strongly influence general relief, as on the Great Plains and the Colorado Plateaus, with their level horizons and flat-topped hills, folds and faults may determine some local features. In uplands such as the Appalachians, upfolds and downfolds have guided the cutting of ridges and valleys. Folding has raised domes out of which peaks such as the Black Hills have been carved. Downwarping may account for the existence of great troughs in which major rivers develop drainage systems. Upwarping and faulting have converted some offshore bottoms into coastal plains. Faulting, likewise, has raised some crustal blocks to form mountains and sunk others to make basins; it has worked strange tricks on drainage and is responsible both directly and indirectly for hosts of irregularities in relief.

Continents and ocean bottoms themselves are the consequences of diastrophism. They are where they are because of crustal balancing. Earlier we saw that the continental blocks consist mainly of relatively light material—granitic rocks patched with sedimentaries—"floating" in the heavier basaltic foundation of the crust, much as icebergs float in the ocean. Readjustments between land masses and the ocean basins have been occurring presumably since the lands originated, and presumably will continue as long as land and sea exist.

Areas of extreme crustal unrest are zones of weakness, marked generally by young mountains and active volcanoes. Typical are the great ridge and rift in the mid-Atlantic, the Eurasian mountain chain running from the Alps to the Himalayas, and the Pacific coasts of the Americas with their volcanoes, systems of highly active faults, and relatively new coastal mountains. Here are found numerous direct evidences of folding and faulting in recent time. But other parts of the continents, now relatively quiet, were likewise active in the past. In North America the most active zones 250 to 200 million years ago were in the Appalachian region, and about 70 to 20 million years ago in the Rockies. Disturbances still occur in the vicinity of these mountains, especially the Rockies, but most of the relief produced there directly by diastrophic activity has been modified by erosion. In the East, likewise, few landforms shaped directly by folding and faulting sur-

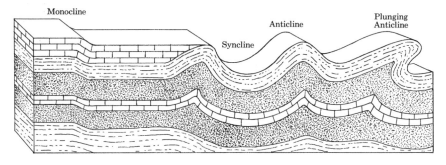

Types of folds: Many variations exist on these basic types. Actual folds occur with varying degrees of distortion and often are broken by faults.

vive. Here as in the Rockies the details of our landscapes, though conditioned partly by the ancient displacements and distortions of rock, are primarily erosional.

Fold Patterns

Despite its seeming rigidity, rock can be deformed without breaking if enough force is applied long enough under the right conditions. The force can involve tension, or pulling apart; compression, or squeezing; torque, or twisting; or a couple, in which two forces in the same plane act in opposite directions. But if the rock is not to fracture, a powerful confining pressure is necessary. One mile deep in the crust the pressure may be as great as six tons per square inch. Under such stress rock can be slightly plastic, so that when a deforming force is applied rock grains shift their positions, mineral crystals distort along their structural planes, or the atoms and molecules reorganize to form new minerals in the process called metamorphism.

In a fold the rock on the inside is compressed and the rock on the outside is put under tension. The rock's density is increased on the inside, decreased on the outside. Particles may move to equalize the density again, especially if the rock is impregnated with fluids that can act as lubricants. If the rock is sedimentary, a bending force may cause the strata to slip along the bedding planes, much as individual pages of a paperback book slip when the book is rolled. If metamorphism is involved, the newly formed minerals may crystallize into patterns corresponding to the directions of deformation.

The axis of a fold (visualized as a longitudinal center line along the top of the crest or the bottom of the trough) may be wavy or undulating. In cross section the trough may be sharp-pointed or rounded, the limbs steeply or gently inclined. The limbs may spread to form a wide-open fold or draw close to form a tighter structure. All the patterns that can be

Chevron folds: These are miniatures found in rock outcrops on the Fleurien Peninsula, south of Adelaide, Australia.

produced by rumpling a piece of cloth have been found in rock strata, and folds can vary from the size of a fingernail to the size of a mountain.

The two basic folds are the anticline, or upfold, and the syncline, or downfold, but variations are unlimited. Some folds are symmetrical; that is, the cross section can be divided into two ilke halves by drawing a line down the middle; but most are asymmetrical, leaning one way or the other. The limbs of folds are themselves sometimes folded, so that we have folds within folds. Some folds have been completely overturned, and still others are recumbent on adjacent strata. Many anticlines if they are followed far enough can be seen to plunge, or die downward, at their ends. Plunging anticlines grade into dome forms, and synclines grade into the forms of basins.

A local steepening of dip in otherwise horizontal or gently dipping beds is called a monocline. The term homocline is applied to strata that dip uniformly in one direction. Often a homocline is simply the remnant of one limb of an eroded anticline or syncline.

A warp in rock strata is what the word implies: a distortion involving both bending and twisting. Sedimentary rock usually shows at least slight warping if it is not actually folded; and folding ordinarily involves some warping.

The monocline in Mount Timpanogos, Utah: Some folding accompanied the faulting that raised this mountain with the rest of the Wasatch Range.

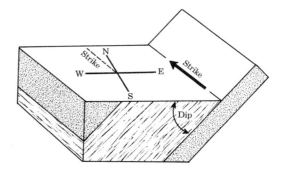

Dip and strike: These are the properties by which the attitudes of rock strata can be precisely described.

Folding and warping involve either uplift or subsidence, or both. A gentle fold can take the form of an arch or a dome, a trough or a basin. Where folding is intense, upfolds and downfolds alternate but the entire complex is uplifted to form so-called folded mountains. This uplift may involve the subsidence of an adjacent terrain, because the accumulation of materials to raise the relief in one place must come from somewhere else.

To describe the attitudes of inclined strata geologists use the terms dip and strike. The dip of a stratum is the angle it makes with a horizontal plane; thus the dip of a stratum thrust up into a vertical position would be 90 degrees. A dip slope is a slope that follows the dip of the stratum beneath it.

The strike of a stratum is the compass direction of the line formed by the intersection of the bedding plane (surface of the rock layer) with a horizontal plane. If such a line pointed 7 degrees east of north, for example, that would be the strike. The term strike is often used to designate valleys that run parallel to the axes of folds; these are called longitudinal valleys also.

Landscapes on Folds

Where rock is buried deep under sediments, as it is likely to be on lowlands, its structure cannot guide erosion, and the relief will express structure only in a broad general way if at all. By contrast, where rock is bared to erosion, as it is usually on highlands, rock destruction is efficient and swift, and structures may be expressed in fine detail.

Erosion of gently inclined strata produces the escarpments known as cuestas. Such an escarpment terminates a long, gradual upward slope. It is maintained by the relatively rapid erosion of weak rock exposed beneath an outcrop of stronger rock. On terrains such as the Colorado Plateaus and the Atlantic and Gulf coastal plains, cuestas appear as straight or curving cliffs or slopes extending scores of miles across country.

If nearly horizontal strata are strongly uplifted, as in the Colorado Plateaus, erosion accelerates and variations in rock resistance become more potent in determining relief. Relatively low areas may undulate according to rock structures, but the major relief consists of tablelands cut out of the uplifted land mass by streams usually along lines of rock weakness. Variations in resistance of the strata sometimes make the sides of tablelands craggy.

Where uplift has been strong and distortion intense, variations in rock strength may produce relief that is very irregular. Greater uplift usually means faster erosion, and faster erosion means more thorough exploitation of lines of weakness. As folds rise, erosion keeps shaving them and exposing the inner rock structures successively. The steeper the limbs of the truncated folds, the more easily water can penetrate between the strata and the more sharply variations in resistance are expressed. Whereas one might expect to see on folds a simple pattern of minor streams running down the sides of anticlines to join rivers following the troughs of synclines, one sees just as often streams running down the sides of synclines to join rivers in anticlines—because it is the lines of weakness, not anticlines and synclines as such, that determine the ridge-and-valley patterns.

The newer Folded Appalachians, usually called "the Appalachians," are classics of folded mountains. Raised between 250 and 200 million years ago, they have been eroded low, buried in rock waste, uplifted again, and subjected to a new cycle of erosion several times. In this process some two or three miles of the original folds has been planed off, and erosion guided by lines of weakness has created the relief seen today. Much of this topography is "inverted." Again and again, weak rock between limbs of a

A cuesta in eastern New York: Strong conglomerate strata, dipping gently westward, form the abrupt eastern scarp of the Shawangunk Mountains.

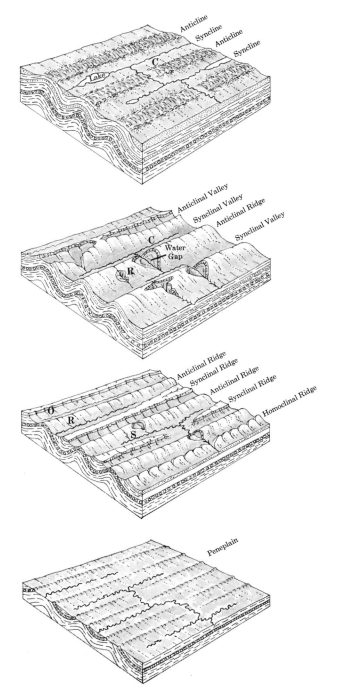

Erosion on folds: As a topography evolves, watercourses are developed along lines of weak rock. (C= consequent stream; R= resequent; O= obsequent; S= subsequent.)

truncated anticline has become the course of a valley, and strong rock between limbs of syncline has become a ridge. Thus, by variations in rock resistance, upfolds have become lowlands, and downfolds, highlands. These inversions may be encountered as often as the "normal" patterns.

A clear-cut example of inversion is provided by Schunemunk Mountain, which rises west of the Hudson Highlands in southeastern New York. Its north portion is the trough-shaped remnant of a syncline. While surrounding shales and limestones have been reduced to lowlands, the remaining parts of the syncline's limbs, which are of strong conglomerate and sandstone, have survived as ridges with a relief of about eight hundred feet. Weathering, stream work, and glaciation have developed a deep longitudinal valley in the weak shale between the ridges. Thus the syncline stands today like a trough set on a platform.

Seen from the air, the Appalachian ridges and valleys have a fairly regular trend controlled by the major folds, though much faulting occurred in the south, especially in the vicinity of western North Carolina. Plunging anticlines and synclines, arranged like long canoes in echelon, make a zigzag pattern which is particularly pronounced in Pennsylvania. The ridges have developed upon strong rock, usually conglomerates, quartzites, and sandstones, and are separated by strike (longitudinal) valleys following weak belts, usually of shale or limestone. Large valleys usually follow the axes of anticlines and synclines, and lesser ones are in weak rocks along the flanks. Obsequent and resequent streams run down the slopes and occasionally a superposed river cuts across the folds.

Along the flanks of folds, and paralleling them, there are often homoclinal ridges. These are stumps of folds whose upper portions have been eroded off. The ridges formed by stumps of anticlinal strata grow older as one moves toward the center of the structure, and ridges formed by stumps of synclinal strata grow younger. The highest of the ridges are those made by the edges of the most resistant strata.

Homoclinal ridges that are narrow, with rugged profiles, are generally called hogbacks. Ridges far from the main structure often have a long, gentle backslope and thus grade into the cuesta form.

Folded structures are eventually cut low and partly buried in erosion debris. If the region later rises, rejuvenated streams then work to exhume the buried structure. Again the belts of weak rock are exploited and strong relief develops. Such, in part, has been the erosional history of the Appalachians and folded portions of the Rockies.

Domes and Arches

Flying low over western flatlands, such as the Colorado Plateau, one occasionally glimpses a group of ridges that circle a basin or, less commonly, a central hill. Such are the ruins of domes. The ridges are the edges

Folds in the MacDonnel Ranges, Northern Territory, Australia: Low parallel scarps made by erosion of these folds stretch virtually unbroken for 150 miles.

of upturned strata, and the basin or hill in the center represents either a core of relatively weak rock that has been eroded away or a core of strong rock that has survived.

Such domes are common on sedimentary strata in regions of former volcanism. Usually of the order of a mile or two in diameter, they are formed by the upbulging of rock strata that occurs as rising magma squeezes between them to form the structure called a laccolith. A few of these domes still have the sedimentary cover; in many the basaltic igneous core has been exposed and left as a highland—Navajo Mountain, in Utah, for example. In some domes the core is still hidden beneath a central basin.

Domes of greater size are formed by the uplift of batholiths. These are

reservoirs of granite or other acidic rock that originate at great depth. All known batholithic domes are very old, and apparently their rise has been very slow. Some uplifted batholiths, such as those beneath the sedimentaries of the low Zuni Uplift in New Mexico and the Lake District in England, still have the sedimentary cover. Domes of greater height and, perhaps, of greater antiquity lost their cover millions of years ago, and erosion has been at work on the igneous core ever since. These ancient structures are characterized by very strong rock which, though deeply dissected by erosion, survives today as highlands, such as the Black Hills and the Adirondacks.

Very low, broad domes, some over one hundred miles wide, are common on sedimentary strata. They show no igneous core, but are identified

A laccolithic dome on the Colorado Plateau: Erosion of this structure at Sinclair, Carbon County, Wyoming, has left hogbacks surrounding the central core.

by the cuestas around their edges. Some are better called arches and swells, such as the Cincinnati Arch, which is centered on the Blue Grass region of Kentucky, and the San Rafael Swell of Utah. The state of Florida is formed by a long, low limestone arch—essentially an anticline with a northwest-to-southeast axis.

The Weald, in southeast England, is the western part of an elongated dome (one of several) that once reached from south of the present site of London eastward into the Boulognnais region of France. The dome's excavated central portion now lies beneath the Straits of Dover. The Weald, about 90 miles long east to west and 50 miles wide, consists of a central basin containing a cluster of high hills. The basin, hollowed out of an ancient sandstone, is rimmed by upturned younger strata, with a chalky limestone as the outermost ring. The latter forms the renowned white cliffs of the North Downs and South Downs, and of Dover and Beachy Head. Western Heights, the highest elevations of the Weald, are formed in the northwest by sandstone.

Across the Straits of Dover, on the French side, the other end of the dome has been likewise cut out by erosion. Southwest of the Weald is the chalk Isle of Wight, part of another, smaller dome now mostly covered by the English Channel.

The dome from which the Weald was carved may once have been several thousand feet higher than today's Western Heights. The Weald's rivers, which are continuing their task of dissection, long ago established the drainage pattern that is familiar on domes of sedimentary strata:

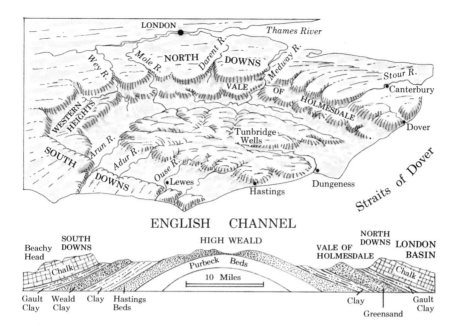

The Weald: Erosion has truncated this dome in southern England, exposing weak rocks around the center. The inner lowland is on these rocks.

streams running down the sides of the dome, fed by streams joining them almost at right angles, where beds of differing resistance overlap.

The erosion of some low domes has reversed the topography. The Nashville Basin south of Nashville, Tennessee, originated as an elongated dome about 60 by 120 miles in extent. Today it is a basin, with a depth as great as 400 feet. The escarpments ringing the basin are formed by the edges of sandstone and limestone strata that formerly extended over the top.

The salt domes of Texas and Louisiana, averaging a mile or so in diameter and a hundred feet in relief, originate because of differences in density between rock masses. Deeply buried salt, the residue of ancient seas, ascends in the crust because of its lightness, making the surface strata bulge up and sometimes breaking through them. These domes often are identified by the saltiness of springs (salt licks) issuing from their sides and

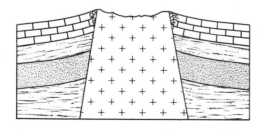

A salt dome: This one has broken through the surface. Many are completely underground.

also by the saltiness of wells drilled into them. Usually the drainage system is radial from the summit, with tributary streams ringing the dome as they follow exposures of weak rock. Some of these structures are capped with anhydrites, gypsum, or limestone; in others the capping has been stripped off and the dome virtually leveled. Salt domes are familiar along the American Gulf Coast, in the Harz Mountains of Germany, and in Rumania, western Russia, the Near East, and northern Africa.

The San Andreas fault: In this aerial view a relative displacement of the relief is visible along the fault. Note also the pronounced derangement of the streams.

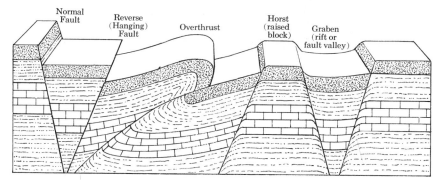

Fault patterns: The displacements indicated here are mainly in the vertical. Actual faults often involve horizontal displacements also, and may be involved with folds.

Fault Patterns

A fault is a fracture that involves the displacement of one body of bedrock with respect to an adjacent one. Major faults occur along master joints—that is, joints of regional extent—but minor faults may occur also along planes of relative weakness. These planes are determined by points at which the rock grains are less firmly joined by cementing materials or where the mineral crystals are weakly interlocked. At their smallest, faults are microscopic; at their greatest they are hundreds of miles long and as deep as Earth's mantle.

The displacements may be vertical, horizontal, rotary, or any combination of these. Usually the fracture plane is off the vertical; then the lower surface of the rock mass above the fracture is called the hanging wall, and the surface facing it is the footwall. If the hanging wall has sunk relative to the footwall, a gravity fault, or normal fault, has occurred. If the hanging wall has been pushed up over the footwall, a thrust fault is the result. Where strata of the hanging wall have actually surmounted the footwall and slid over it, there is a so-called overthrust. Most normal faults are rather high-angle faults, whereas most reverse faults are of the low-angle type. Part of an overthrust block that has become isolated from the main thrust sheet is called a klippe (German, "cliff").

A fracture may appear as a clean crack or fissure, or it may be a zone filled with broken rock (fault breccia) or pulverized rock (fault gouge). Displacement along the fracture is usually called slip.

Since fractures may occur abruptly as a rock mass gives way under stress, an extensive break usually produces an earthquake. The shaking motion can be terrifying and devastating, but displacements that can result from a single disturbance are minor. In the Alaskan earthquake of 1964—the most violent earthquake ever recorded by seismic techniques—the maximum displacement, as measured on Montague Island, was less than forty

A roadside fault, Sandia Mountains, New Mexico: The strata at the left have been displaced upward with respect to adjacent strata.

Slickenslides in a road cut: These grooves in Hudson Highlands gneiss were produced by faulting in the remote geologic past. Exposed recently by highway construction, they will now be subject to obliteration by weathering.

feet (see page 126). In the San Francisco earthquake of 1906 the two sides of the San Andreas Fault were displaced horizontally by no more than twenty-one feet, as indicated by broken fences that straddled the fault. Clearly, mountains and valleys are not made by sudden cataclysms. Mountains miles high and grabens thousands of feet deep form by increments of a few inches or feet over thousands of years.

Landscapes on Recent Faults

Faulting usually goes with strong folding, as in the Appalachians, but it can be severe also where folding is slight, as in the Basin and Range Province. It creates some relief forms directly, and fault zones guide erosion long after these original forms have been removed.

The common relief feature produced by faulting is the cliff or slope known as a fault scarp. If the fault is normal, the exposed part of the footwall becomes the scarp. In a reverse fault the scarp originates as the underside of the hanging wall; but the overhang is eroded away rapidly, so that the scarp superficially resembles that of a normal fault.

A scarp that has not been long exposed to the elements may often be identified by grooves and scratches, called slickensides, on its face. These marks are caused by abrasion during the process of displacement. Rarely, slickensides include very small, sharp "steps" at right angles to the direction of movement. The direction is always down the steps, not up.

Most scarps result from the tilting of a crustal block. The Wasatch Range of Utah was raised in this way, and so was the block from which have been carved the Sierra Nevada of California. Whereas the steep scarp of the Wasatch block faces west, that of the Sierra block faces east. Both blocks sank at one end while rising at the other.

Any depression due to faulting is a graben. The sides of grabens are fault scarps. In some, such as Death Valley in California, the bottom is the top of a block that sank. Grabens along the Great Rift Valley of eastern Africa show evidence that the flanking blocks were pushed partly over the sunken ones; hence these grabens have sometimes been called ramp valleys. Grabens are very irregular in form—rarely like the neat diagrams seen in textbooks.

Deep grabens of recent origin often make dramatic scenery. Death Valley is one of the unforgettable sights in the American West. In Germany the most picturesque part of the Rhine River's course, in the Black Forest region, is a graben. Scotland has its spectacular Midland Valley, and Russia the graben that has become Lake Baikal. Africa's Rift Valley contains great lakes such as Nyasa and Albert, majestic waterfalls like Victoria and Kalambo, and a number of volcanoes as well.

The opposite of a graben is a horst—a fault-bounded block heaved up

A horst in the Mediterranean: Sifnos Island, like others of the Greek group, is an uplifted limestone fault block. Fault scarps form its sides.

by some disturbance beneath it or squeezed up by adjacent blocks thrusting against it. Among the largest known horsts are those of the Vosges Mountains in France and the neighboring Black Forest Plateau in Germany. Others flank the Dead Sea on the east and west. The islands of Greece are horsts which, judging from the elevations of old beaches on or near their tops, have been raised as much as a mile. The steep sides of the islands are fault scarps, at which the horizontal strata end abruptly.

A rift is a crack or fissure formed where a fault plane intersects the ground surface. Usually it appears as a long, narrow zone of broken rock that shows signs of displacement—such signs as intermittent low scarps on either side, or sags, which often are filled with water. Horizontal displacements, if any, may be indicated by opposing scarps that are offset, or by diverted streams or offset stream channels, as in the grabens of the Los Angeles Basin in California.

Faults are often intricately involved with folds. Long-established folds become disorganized by faulting. In the usual evolution of folded moun-

Mountains in a region of thrust-faulting: The tilt of strata in Mount Rundle in Banff National Park, Alberta, strongly determines the profile.

tains, the folding tends to predominate at first and faulting later. Distortions and dislocations of strata in the Alps have made such a jumble that the piecing together of Alpine history is still going on after a century.

Strong relief due to faulting is slow to develop, and during intervals of stillstand a scarp may be practically obliterated. Even if relief develops faster than erosion can destroy it, the amount of relief at any time will never equal the total displacement. On the west side of the Wasatch Range, for example, the vertical displacement has totaled more than three miles, the actual relief probably never much exceeded a mile, and the relief today is even less than that.

In fault topography we see again and again a familiar structure: a tilted block with a steep foreslope and a gentle backslope. If the tilting has been intermittent, as is usual, the foreslope may show a series of steplike slopes steepening toward the bottom, each representing a phase of uplift. In a thrust fault the front of the block may overlap the strata against which it has been pushed, and the pushing may have caused folding.

On a young fault scarp the gullies and ravines are the surest indicators of the scarp's origin. They are poorly organized and perhaps shallow. Whereas on ordinary slopes the stream channels widen toward the bottom, on a young fault scarp they narrow, because there has been little time for cutting near the bottom. Toward the bottom the narrowing may result in an hourglass shape, and if uplift has occurred very recently, the gullies may end as hanging valleys above the base of the scarp.

The base of a young scarp is on the fault line—it has not retreated yet under the assaults of erosion. Erosion has not yet removed the zigzags, if any, in the fault line caused by variations in rock strength. Talus has not accumulated to great depth, slickensides may be visible, and little fault scarps may be seen along the foot of the main one. If uplift has recently disturbed a stream in the vicinity, diversion or ponding may be noticed.

As the typical V-type ravines are developed in the face of the scarp, more and more of the original surface is removed. Meeting at the top, the V's make a profile that suggests sawteeth, and the face of the scarp is cut into triangular facets, with the points at the top, like those seen on the west side of the Wasatch Range. As the V's continue to widen, the remainder of the original scarp face is destroyed and the base of the cliff retreats from the fault line. Meanwhile the relief produced by the uplift is gradually cut down and softened. If the caprock on the fault block is strong it may

A fault scarp: The walls of Death Valley still show triangular facets and hourglass valleys, typical of a fault surface not yet fully destroyed by erosion.

last a long time, and since it corresponds to the rock on the surface of the surrounding land, it stands as proof that the block was indeed uplifted. But even such rock is destroyed in time, and barring a renewal of uplift, the scarp is reduced to a low ruin.

Highlands and valleys of fault origin are seen only where the faulting has been recent, geologically speaking. American examples include the Basin and Range Province, the Coast Ranges, and the Andes. In Eurasia the active regions are along the margins of the Alps, the Carpathians, and the Himalayas, and in Africa activity is centered in the northwest around the Atlas Mountains, and in the east along the Rift Valley. Zones of disturbance include also the volcanic island arcs running from Alaska and the Aleutians down the western Pacific to the South Seas, and the arcs of the Atlantic, particularly the West Indies. Displacements are occurring elsewhere too, as minor earthquakes testify, but in relatively quiet areas they are not producing any visible relief.

Since erosion is relatively rapid in humid regions, it is on arid terrains that fault-produced landforms are longest-lived. In the United States the main showplace is the Basin and Range Province. On its eastern side is the Wasatch Range, formed by a tilted block with a west-facing scarp, and on the western side is the Sierra Nevada, with its east-facing scarp. In the southern reaches of the province, in the vicinity of the Mexican border, the

A klippe in Glacier National Park: The great bulk of Chief Mountain lies on rocks of younger geological age and thus is known to have been moved from its place of origin.

fault-block highlands are old and much reduced, and drowning in rock waste, but to the north the scarps are younger. The highlands of Idaho and Montana and the Klamath Lake region of northern California and Oregon have many newly raised fault blocks with fresh scarps.

One of the most-discussed fault features in North America is the Lewis Overthrust in Montana. Here a giant block about three hundred miles wide was pushed over an adjacent one to a distance of forty miles. Chief Mountain, in northwestern Montana, is an isolated remnant of the overthrust block—that is, a klippe. It is identified as such by the fact that it is a mass

The Montague Island fault in Alaska: In the earthquake of 1964 the terrain to the left of the fault line was raised.

of Precambrian rock lying on Cretaceous rock, which is much younger. In the Eastern states, comparable low-angle thrust faults with displacements of twenty to thirty miles have been identified in the southern Appalachians.

Landscapes on Old Faults

As erosion destroys relief directly produced by faulting, it creates a new relief. This is influenced by displacements due to the original faulting, since these altered the attitudes and the weak-strong relationships of rock masses at depth. As erosion proceeds, it bares new weak-strong patterns and the relief changes accordingly. It may change most radically if the rocks are strongly tilted sedimentaries, because of the variations in strength from stratum to stratum.

As the original relief is destroyed, erosion may create an inverted topography on faults much as it does on folds. Visualize, for example, a flat terrain with a thin, strong top stratum overlying a thick but weak one. Faulting here produces a horst, which is subjected to vigorous erosion. When the hard cap of the horst has been destroyed, the remainder is quickly reduced to the level of the surrounding land. But now the site of the horst is a zone of weak rock surrounded by strong rock. If the drainage is good, erosion in this zone will be relatively rapid and make a lowland.

This example demonstrates that as erosion exposes weak and strong rocks successively on opposite sides of a fault plane, scarps may be developed on either side of the plane. The scarp always forms on the side where the rock is stronger. With time, erosion may produce a succession of scarps facing first one way and then the other.

Clearly, then, there are two kinds of fault scarps: true fault scarps, produced directly by faulting, and so-called fault-line scarps, produced by differential erosion along a fault line. The true fault scarp is always on the upthrow side of the fault plane, but the fault-line scarp can be on either side.

A fault-line scarp is called resequent if on the upthrow side, so that it faces in the same direction as the original fault scarp. If on the downthrow side, facing the other way, it is called obsequent. All fault-line scarps are either resequent or obsequent.

Scarps created by a combination of faulting and erosion are called composite. The great east-facing scarp of the Teton Range of Wyoming is believed to have been formed originally by the tilting of a fault block during the Pliocene Epoch, a few million years ago. Since then, erosion has cut away weak rock east of the fault faster than the crystalline mountain rock west of it. Thus the upper portion of the mountain slope is a fault scarp, and the lower portion a resequent fault-line scarp.

Well known as a resequent fault-line scarp is the Ramapo Front, an

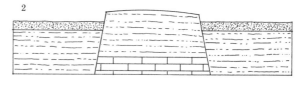

Fault scarps and fault-line scarps: Faulting originally produces fault scarps (*top*). Erosion destroys these (*middle*) and eventually produces fault-line scarps (*bottom*), reversing the topography.

800-foot-high escarpment which forms the southeastern edge of the Ramapo Mountains and Hudson Highlands of northern New Jersey and southeastern New York. The uplands, consisting of strong granitic rocks, are the remnant of an ancient horst raised above a terrain of weak sedimentary rocks. The original highlands that were sculptured from the horst were destroyed long ago, but a scarp is still maintained on the highland side of the fault line because the rock there is stronger than the rock of the adjacent lowland.

Obsequent fault-line scarps can be seen in the Lebanon region of southeastern Pennsylvania. Terrain to the west of the original fault was uplifted, but its rock was weak and erosion has converted the fault-produced highland into a valley. East of the fault the terrain was depressed, but the strong rock there has resisted erosion so well that it now forms the Piedmont Upland.

Some of the most impressive fault-line scarps of the Eastern states are encountered in the Adirondacks of New York, especially in the region of the higher peaks just west of Lake Champlain. Here, during the very distant past, faulting was extreme, and the old displacements have dramatically influenced the more recent shaping of the landscapes by erosion. Many of the depressions, such as those occupied by Saranac Lake and Lake George, are bounded by fault-line scarps.

In North America the most extensive of the old faulted, deeply eroded terrains are seen among the Old Appalachians—the Great Smokies and the Blue Ridge, the New England Upland with the Ramapo Mountains and

A fault-line scarp: The view is along the Ramapo Front, at the southeastern edge of the Hudson Highlands. The highland is of granitic rock; the lowland, of sandstone.

Hudson Highlands, the Berkshires and Green Mountains, and the Laurentian Hills of southeastern Canada. These cores of ancient folded ranges have been severely faulted during several episodes of mountain building, and have been buried and exhumed again. Little remains of their original relief, yet among the distorted, broken crystalline rocks the work of the ancient forces can be traced. Streams have cut ravines along old fracture lines. Some of the valleys are undoubtedly grabens. Many cliff faces represent joint or fault planes from which large blocks have drifted or slid downslope. Some fracture zones are full of fault breccia, and others were injected millions of years ago with hot fluids that cooled to form assorted minerals, including much quartz. Rock faces long bared to the weather have lost the marks of faulting, but those newly exposed in highway cuts show slickensides. These highlands are showplaces of distorted bedrock denuded by erosion over a time span which, despite all our techniques of measurement, the human mind cannot really comprehend.

VI

A Gallery of Mountains

Carved from a fault block: The Grand Tetons of Wyoming demonstrate the abrupt rise typical of fault-block mountains. They show also the results of alpine glaciation.

The supreme expression of folding and faulting is mountains. These come into view always as a surprise, an anomaly of the landscape, a special creation—rising suddenly, massively, even fiercely out of ordinary land. Primitive peoples have regarded mountains as separate worlds, the abodes of the gods, and almost the same feeling survives today among those who tramp the high places and, with rope and piton, set their faces toward the peaks. The classical Greeks showed respect for mountains by attributing them to the pull of the stars. When God in the Bible spoke to man, it was from a mountain. Today the heights are emptied of such superior beings, but they are viewed with hardly less wonder as the most spectacular places on earth.

Mountains have long been a puzzle. Why should they rise so high, so abruptly? The ancients wondered at the enormous forces evidenced by the folding and fracturing visible in mountain rock. The presence of marine fossils in strata miles above the sea was long a tantalizing paradox. Our modern awareness of mountains as phenomena that express crustal readjustments is barely a century old. Exactly how mountains do originate is still as much a matter of speculation as of known fact.

James Hutton, the Scottish geologist, swept away most of the old ideas about mountain origins with his assertion that Earth's crust was shaped in the past by the natural forces that are shaping it today. After Hutton the next important concept came from the American James Hall (1811-98), who pursued his great study of the Appalachian Mountains for more than a half century. These mountains, Hall discovered, consist of erosion sediments that originally accumulated to depths as great as seven or eight miles and then, having changed to rock, were folded and lifted in a slow, gigantic spasm of the continent. Hall saw that so much sediment could accumulate only if the crust sank beneath it, but he was perplexed by the question of how such a mass could be raised to make mountains. An explanation was offered by another American, Clarence Dutton (1841-1912), in his theory of isostasy. Mountains, said Dutton, can be raised by crustal balancing processes, and he pointed out that mountain ranges generally consist of rock lighter—as gravity measurements show—than rock underlying them.

With the twentieth century came a knowledge of Earth's interior heat and the concept of great convection currents in the mantle. These, along with upsets of balance caused by transfers of erosion sediments, appeared to go a long way toward explaining mountain formation in general terms at least.

The Birth of Mountains

According to current theory, mountains originate in long, broad troughs called geosynclines. These troughs, hundreds or thousands of miles long, result from downwarping. It is thought that where opposing currents in the mantle turn downward beneath the crust they may drag the crust down with them to form the geosyncline. Another cause of downwarping might be the contraction and expansion of crustal materials due to changes of state (for example, liquid to solid) or chemical reactions. Again, it seems likely that as erosion debris is swept down onto lowlands, crustal balance is disturbed, as Dutton said, and compensating shifts of crustal blocks result.

Whatever their origin, geosynclines do become the dumping grounds for vast volumes of sediments, both erosional and organic. Over millions of years the sediments, gradually turning to rock, accumulate to depths of miles, and the bottom of the geosyncline sinks. Lateral forces, perhaps of convectional origin, begin pushing inward from the sides of the trough, folding and breaking the strata. Fierce heat at depth melts and meta-

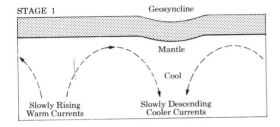
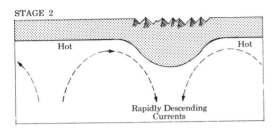
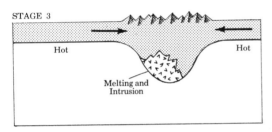

Evolution of a geosyncline: Mountain building is due not only to the effects of Earth's heat, as indicated here, but to other causes including balancing processes in the crust.

morphoses the sediments. Faulting allows magma to squeeze upward from the mantle into the crust, and some of this erupts to form volcanoes. Meanwhile a general uplift begins, and in the course of more millions of years the folds become mountains.

During active mountain building, uplift is faster than erosion, and the mountains keep gaining in elevation. But they never achieve a height equal to the total uplift. In the Appalachian geosyncline, sediments in the vicinity of what is now North Carolina were at least eight miles deep, yet geologists find evidence that the mountains there never attained an elevation much exceeding three miles.

The continents are said to "float" in the basaltic underlayer of the crust because they consist of lighter materials. It is for the same reason, apparently, that mountains stand high on the continents. Mountain roots, which consist of granite-related rock, plunge deep into the basaltic layer. Mountains "float" in the continents and are no more a load on them than icebergs are a load on the ocean. But floating is not a static condition. Crustal forces continue to disturb mountain regions long after the original uplifts, and as erosion debris is swept down onto the lowlands, new uplifts may occur to restore the balance. A mountain region remains a zone of instability as long as the mountains have roots and, therefore, the elevations are substantial. But in the length of ages every mountain range, roots and all, does succumb to the elements and the region goes quiet—while new mountains rise elsewhere.

The continental shields, which are platforms of much-metamorphosed, much-distorted rocks approximately centered in the continents and partly covered by younger strata, appear to be the roots of mountains of the past. Radioactivity measurements indicate that these rocks are two to three billion years old. The formations beneath them are still older. Around the edges of the shields are younger metamorphics which, too, appear to be highland remnants. The evidence is strong that mountain building over the eons has been a common event, fundamental in the long-term evolution of continents.

Mountain building, broad uplifts, and volcanism together have occurred intermittently. Periods of maximum disturbance are, as we have seen, convenient markers for dividing the geologic past into eras and periods. The rise of highlands and the eruptions of volcanoes have altered atmospheric conditions and thus influenced the fortunes of animal and plant species. The Geologic Time Table shows the rise of the newer Folded Appalachians as ending the Paleozoic Era; the uplift of the Andes, the Alps, and the Rockies as ending the Mesozoic; and the elevation of the newer Alps and Himalayas as marking recent stages—possibly the closing ones —of the Cenozoic. Our age of rugged scenery—scenery more rugged than during most of the geologic past—is featured most of all by mountains.

Every continent today has its mountains. They range from the uncounted hills to the great massifs, some ancient and worn low like the Laurentians of southeastern Canada, others middle-aged and substantial like the Appalachians, and a few still young, very steeply rising, like the Himalayas. The great ranges make a discernible pattern, beginning as the Andes at the southern tip of South America, running northward into Central America, continuing as the Coast Ranges and Aleutians in North America, and then swinging down the western Pacific along an axis marked by the Kurile Islands, Japan, the Philippines, and Indonesia. From here the chain runs west into south-central Asia, achieving its greatest heights in the Himalayas, and continues through the Middle East to the Caucasus. While one branch reaches down through East Africa, the main line continues westward, becoming the Alps and the Pyrenees of southern Europe and the Atlas Mountains of northwest Africa.

Most of these ranges are young- to middle-aged. The young volcanic group appear to be currently involved in land-building along continental margins. Somewhat older ranges, such as the Rockies and the Alps, were involved in earlier continental growth. Mountain masses not included in the pattern just traced, such as the Appalachians, the Urals, and the Central Asian highlands, date from a still remoter past.

Some massifs consist predominantly of folds. Such ranges, which often achieve great dimensions, characteristically have a granitic core formed by deep crustal heating before or during uplift. Associated with folded structures there are usually fault-block mountains, less massive but often more ruggedly scenic. Some block mountains consist of the broken crystalline cores of old folded highlands.

Each category of mountains has its peculiar features. Folded ranges may have parallel ridges and valleys, cut according to lines of weakness in sedimentary strata, with ridges often running long distances without interruption. Block mountains commonly are more angular, more irregular. Upon them erosion may follow lines due to faulting, so that the profiles often directly show patterns of displacement. Volcanoes generally feature steep-sided cones or low, broad domes, according to the nature of their lavas.

The ruggedness of mountains is due as much to degradation processes as to diastrophism. High elevation means swift downcutting by streams, and swift downcutting makes steep-sided ravines and valleys. Weathering, especially by frost action, is intense, and mass wasting is rapid. As rock

In the Appalachian Mountains of West Virginia: The range here is seen from Cave Mountain, south of Petersburg. Note the cliff-forming tendency of the inclined strata.

disintegrates and joints widen, fragments slide, tumble, and fall to lower levels, bringing down other fragments with them, and leaving fresh surfaces exposed to the elements. The sharpness of relief is greatly enhanced if valley glaciers have been among the sculpturing agents. All in all, it seems that the beauty of mountains is proportional to the rate at which they are being destroyed.

Folded Ranges: The Appalachian Story

North America's major folded mountains today are the newer Folded Appalachians. These make a favorite model because, compared to other folded ranges such as the Alps, they have a relatively regular structure, especially in the north. Their story begins with a chain of volcanoes that grew up out of the Atlantic Ocean during the Precambrian Era, about a billion years ago, along a zone somewhat west of the present East Coast between Newfoundland and Georgia. As the belching cones emerged above the waves, they were attacked by waves and weather, and the debris drifted down into hollows between the islands and into a geosyncline west of them. Meanwhile sediments were cascading into the geosyncline from lands to the west. During the Paleozoic Era, which began about 600 million years ago, the volcanic peaks, which are known as the Ancestral Appalachians, were nearly leveled, and parts of the geosyncline were folded and heaved up, and then cut down again, in several cycles. But along the eastern fringes of the trough violent volcanism persisted. Then, toward the close of the Paleozoic, 250 million years ago, a series of upheavals in the geosyncline produced the newer Folded Appalachians.

As uplift slackened, erosion became dominant. During the ages since then, uplift has occurred intermittently, perhaps partly to compensate for transfers of rock waste to the Atlantic coastal plain. During quiet intervals erosion has reduced extensive areas to peneplains. Upper parts of the folds have been cut off, and the stumps have been buried and exhumed several times. During the Pleistocene, glacial ice shaved and gouged the mountains from northern New Jersey and Pennsylvania northward. Today the Folded Appalachians are being strongly eroded again following the Cenozoic uplift. Most of their former bulk now lies scattered as sand and clay on the Atlantic and Gulf coastal plains and out on the continental shelf.

Appalachian mountain building is represented today by four fairly distinct provinces. Of these the easternmost and in part the oldest is the Piedmont, whose formations underlie the coastal plain and are exposed west of it, from Alabama to eastern Pennsylvania, mostly as low, rounded hills. Some of these hills are the ruins of peaks that supplied the sediments for younger Appalachian highlands. The rocks have been intensely folded, faulted, and metamorphosed; many are granites and altered basalts. Radio-

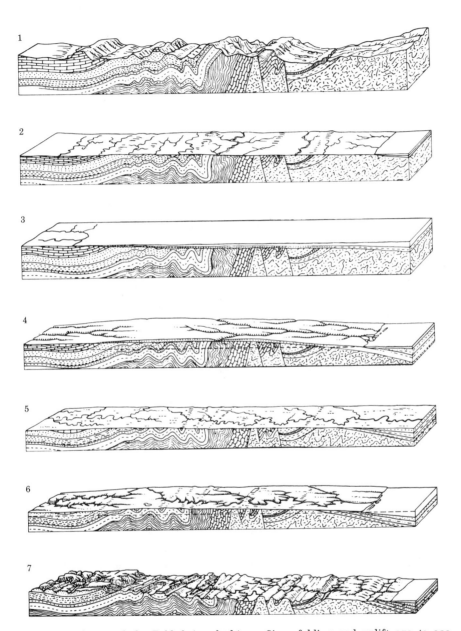

Probable evolution of the Folded Appalachians: Since folding and uplift 250 to 200 million years ago (1), the terrain has been peneplained (2), covered by the sea (3), arched (4), again peneplained (5), arched once more (6), and subjected to the erosion cycle that continues today (7). The section shown reaches from the Appalachian Plateau (*left*) to the Atlantic Ocean at about the latitude of New York City.

Exhumed by erosion: The Blue Ridge of northern Virginia viewed northward from a lookout near the Skyline Drive. Tilted layers of metamorphosed basalt form cliffs.

activity tests indicate ages of around a billion years, but some of the intruded igneous masses are much younger.

Immediately west of the Piedmont is the Blue Ridge Province, with chains of highlands running intermittently from Georgia to southern Pennsylvania. This division takes its name from the Blue Ridge of Virginia but also includes the Great Smokies. The sedimentary strata that originally covered these highlands were eroded off and used in the making of the Folded Appalachians. Only the cores remain—much-faulted, highly altered rocks such as gneiss, granite, schist, and marble.

Counterparts of the Piedmont and the Blue Ridge rise in New Jersey, New York, and New England. Most prominent are the highlands that begin as the Ramapo Mountains of northern New Jersey and southeastern New York, continue as the Hoosacs and Berkshires in western Connecticut and Massachusetts, and appear in Vermont as the Green Mountains. These are the eroded cores of ancient folded ranges. The White Mountains of New

Hampshire are of a different nature, having been carved out of uplifted masses of younger volcanic rock.

West of the Blue Ridge Province and definitely younger is the Ridge and Valley province, which includes the highlands already termed the newer Folded Appalachians. Running in roughly parallel ridges from Alabama to southeastern New York, these have been reduced less than the Blue Ridge and still carry sedimentary strata—limestones, shales and sandstones, and conglomerates, with infolded beds of coal. The rocks have suffered much breakage. Where distortion has been severe, bituminous coal formations have been converted into the more compact anthracites. In the southern sector some broken folds have been thrust at a low angle westward to distances of twenty to thirty miles. Although the original Appalachian folds have been cut low, they can be traced by projecting the stumps, and their sequences can be reconstructed by the use of index fossils and by lithologies, or rock analyses.

The fourth and youngest division of Appalachia is the Appalachian Plateau, which lies west of the Ridge and Valley zone. This much-dissected highland with its horizontal and gently warped rocks is known from Maryland southward as the Cumberland Plateau and from Pennsylvania northward as the Allegheny Plateau. Its eastern edge constitutes the Cumberland

The Front Range in the Southern Rockies: Longs Peak is the highest summit (*near top center*) in this view from Trail Ridge Road in Rocky Mountain National Park.

Escarpment and Allegheny Front, while the western edge forms lesser escarpments in northern Georgia, Tennessee, Kentucky, and Ohio. In the northeast a high scarp reaches up as far as central New York, and out of it have been carved the Pocono Mountains of Pennsylvania and the Catskills of New York.

Folded Ranges: The Rockies

The Rocky Mountains of North America are a potpourri. Among them are folded ranges which are about 200 million years younger than the Folded Appalachians and have been much less reduced by erosion. There are also clusters of volcanic cones, fault-block mountains, and peaks carved out of uplifted batholiths.

During the mid-Mesozoic a geosyncline several thousand miles long developed in what is now the Rocky Mountain region, and this depression was invaded by marine waters to form the Sundance Sea. The seaway at first reached from the Arctic down to Arizona and New Mexico; then between 100 and 75 million years ago it lengthened and split the continent, drowning an area 200 to 700 miles in width—as far east as Lake Superior and as far west as central Montana. Some 50 to 60 million years ago, diastrophic spasms began along the entire geosyncline, and during ensuing millions of years strata formed from a billion cubic miles of sediments were heaved up to elevations of 10,000 to 20,000 feet. Meanwhile intense volcanic activity occurred in and near the zone of uplift.

As this activity waned, erosion became dominant and, by the middle of the Cenozoic, most of the mountains had become stumps nearly buried in debris. Then new uplifts, beginning about 20 million years ago and climaxing in the Pleistocene, started another cycle of erosion. Many of today's rivers in the Rocky Mountain region became superposed during this cycle—the Arkansas on the Front Range, the Snake on the southern Tetons, the Bighorn on the Bighorns. Intense glaciation occurred as far south as New Mexico's higher ranges. Since the Pleistocene, uplift has waned and streams have taken command.

The most massive folded ranges of the Rockies are those running from western Montana northward. In Canada they are the Canadian Rockies, and in Alaska the Brooks Range. Their structures compare with those of the Appalachians but are larger, with summits as high as Mount Robson's 12,972 feet. All these highlands are still capped with sedimentary rocks; no basement formations have been uncovered. The folds are generally narrow and tight, with many strong eastward overthrusts, such as the Lewis.

Most of the Rockies south of Montana are broad-backed, steep-sided anticlines that have been eroded down to their Precambrian crystalline cores. Typical are Wyoming's lofty Wind River massif, reaching 13,785

The Canadian Rockies in Alberta, Canada: Peaks in Waterton Lakes National Park, forming impressive scenery, are faulted and uplifted masses of sedimentary rock.

feet in Gannet Peak, and the Bighorns, rising to 13,165 feet in Cloud Peak. To the east, straddling the border of Wyoming and South Dakota, are the Black Hills, of which Harney Peak, reaching 7242 feet, is the highest summit.

The Uinta Mountains of Utah are unique in the broad-backed anticlinal group because they are carved from an east-west rather than a north-south arch, and because the Precambrian rocks of their summits are sedimentary, not crystalline. Even after losing an estimated 15,000 feet to erosion, the Uintas have a relief exceeding 6000 feet.

The southern Rocky Mountain Province also displays lofty anticlinal ranges. Among these are the Laramie, Medicine Bow, and Park ranges in the north, and the Sawatch and Front ranges to the south. The latter includes Longs Peak and Pikes Peak, with elevations of 14,255 and 14,110 feet respectively. In the Sawatch Range is Mount Elbert, which at 14,431 feet is the highest of all the Rockies in North America. Also numbered among Colorado's big ranges are the intensely folded and much-faulted Sangre de Cristo Mountains, which extend down into New Mexico.

Along the margins of the Rocky Mountain anticlines, sedimentary strata have been steeply tilted. Erosion has cut all these strata low and in some areas has buried them, but elsewhere the stumps still rise as hogback

The vicinity of the Valais Alps, Switzerland: The knife-sharp ridges and horn peaks show the work of vanished glaciers. In the background is the Matterhorn.

foothills. Spectacular tilted strata called "flatirons" are seen on the eastern edge of the Front Range in Colorado. The high, sharp hogbacks in the Garden of the Gods and the Red Rocks near Pikes Peak have made these places famous.

All the northern and central Rockies, and the higher summits down into New Mexico and Arizona, were subjected to the Pleistocene glaciation. Especially in the north there are the characteristic horn peaks, U-shaped valleys, cirques, rock-basin and moraine lakes, hanging valleys, and outwash plains.

From Arizona and New Mexico the American cordillera extends southward through Mexico as the Sierra Madre Oriental in the east and the Sierra Madre Occidental in the west. These ranges are mostly fault-block mountains and volcanoes; volcanoes are common through Central America as well. In South America the cordillera becomes the mighty Andes, extending more than 4000 miles from the northernmost tip of Venezuela to Tierra del Fuego. The Andes are mostly a chain of truncated folded structures with crystalline cores exposed as the summits. Among the many lofty peaks there are, especially in the north, numerous volcanoes. One of these, Argentina's Aconcagua, 22,835 feet high, is the highest mountain in the entire western hemisphere.

Folded Ranges of Eurasia

Somewhat younger than the Rockies is the far-flung Alpine-Himalayan system. This many-branched chain begins in the west with the Pyrenees of France, the Sierra Nevada of Spain, and the Atlas Mountains of northwest Africa. In southern Europe it forms the Apennines of Italy, the Dinarics of Albania and Yugoslavia, and the Pindus Mountains of Greece. Farther east it includes the Carpathians, the Transylvanian Alps, the Taurus Mountains

Erosion on limestone mountains: The rock of Italy's Dolomites formed on a sea bottom. Today it expresses rapid erosion in vertical joints. The scene is at Bolzano.

of Turkey, and the Caucasus ranges east of the Black Sea. Continuing through the rugged highlands of Iran and Afghanistan, this far-sprawling system attains its greatest elevations in the Himalayas of northern India and Nepal.

All these ranges had a common origin in the vast mid-Mesozoic geosyncline that reached from southern Europe eastward as far as Indonesia. Occupying this geosyncline was the Tethys Sea, giant ancestor of today's Mediterranean. Out of its bottom, mountains have been uplifted at intervals during the past 75 million years.

The Alps and the Carpathians, the youngest members of the group, have already undergone several uplifts and some denudation. The Alps particularly are known for a wild complexity of structure, due to extreme folding and twisting, extensive breakage, overturning of folds, and thrusting of broken blocks over adjoining lowlands to distances of thirty and forty miles. Here are many klippen, including the Matterhorn, the Weisshorn, and Mythen Peak. The Alps have been much reduced by erosion, especially Pleistocene glaciation, and recent uplifts have not kept pace.

The Himalayas, by contrast, are still rising rapidly. The southern portions, which are the youngest, have gained an estimated 6000 feet of elevation during the past 600,000 years. The earthquakes that intermittently shake this region testify to the vigor of the forces still at work there.

Fault-block Mountains

Fault-block (or simply "block") mountains often are neighbors to folded ranges and volcanoes, but they appear independently also. Typically they show high-angle normal faults along their flanks; the mountainsides are the footwalls. A single block may form an entire range, usually being tilted so that one side has a steep slope and the other a much gentler slope, as is true of the Sierra Nevada, the Wasatch, and the Tetons. The loftier blocks usually have lost their upper, younger portions to erosion, so that the summits of today consist usually (but not in the case of the Sierra Nevada) of Precambrian crystallines. Most block mountains are extraordinarily irregular and rugged because of structural dislocations.

The Harz and Vosges massifs of Germany and France, respectively, and the islands of Greece have already been mentioned as well-known areas of uplifted fault blocks. In England there are the North Pennines, sculptured from a fault block that dips down toward the North Sea. Central Asia's block mountains include the worn hills of the Gobi Desert and the Tarim Basin, which is a terrain like North America's Basin and Range Province. Africa too has many block mountains in its deserts, notably in Libya, but its greatest specimen is the Ruwenzori horst in the Great Rift region. The top of this block is more than 10,000 feet above the plateau and 16,794 feet above sea level.

In the United States block mountains are scattered over many areas of the West. Here they were raised recently enough so that the original relief has not yet been erased. In the Basin and Range Province, which is bounded by the Sierra Nevada on the west and the Wasatch Range on the east, blocks rise around depressions which are the tops of downfaulted blocks. The intense faulting here obviously was associated with the uplift of the Rockies. The low average elevation of this province may be due to the withdrawal of crustal material at depth to compensate for the uplift.

Toward the south, particularly in the Sonoran Desert of southern Arizona and northern Mexico, the tilted blocks appear in a state of worn old age. Uplift has ceased and erosion rules. Typically the foreslope of the block has lost much of its steepness, its original surface has disappeared, and the base line has retreated far from the fault line. The ravines extend

Erosion on the Paradise Range, Nevada: The original fault scarps have been cut back by erosion. Rock waste lies deep at the foot of the mountains.

ROCK, TIME, AND LANDFORMS · 146

all the way to the foot of the slope, although they still may not widen as do the valleys in slopes of erosional origin. The streams having lost gradient, deep masses of talus and alluvium surround the mountains.

The Wasatch Range in eastern Utah has been carved from the western edge of a tilted block 130 miles long, north and south. The scarp still rises 4000 feet above neighboring Great Salt Lake, and the triangular facets have not yet been removed. Small scarps in the alluvial fans are signs of renewed uplift. Hot springs at the base of the scarps indicate that the faults reach far down to a region of heat. An unusual feature is a very crooked fault line that passes straight through the weak strata but goes around the stronger ones, so that spurs are left on the scarp. The pitch of this western slope has been much reduced, but landslides still occur.

The Sierra Nevada of California are cut from a westward-dipping plutonic block some 400 miles long and 50 to 80 miles wide. This was raised in the early Cenozoic, reduced nearly to base level, and lifted again in the late Cenozoic. Eleven of the summits still stand 14,000 feet or more above the sea, and the eastern scarp rises one to two miles above the Great Basin. Into the western slope streams and Pleistocene valley glaciers have cut profound canyons, notably those of the Yosemite and King's rivers. King's Canyon, with its depth of 8000 feet, is the deepest on the continent; many neighboring canyons have depths of nearly a mile. The eastern slope, much steeper, is being cut up very rapidly by torrential streams, and the erosion debris is forming great alluvial fans on the lowland. At the foot of this eastern scarp are signs of recent uplift—small, fresh scarps with slickensides; hot springs, indicating deep faults that are active; chains of grabens; and the recently active Mono Craters. One of the grabens is occupied by jewel-like Lake Tahoe, which is 22 miles long, 12 miles wide, and more than 1600 feet deep.

West of the Sierra Nevada is the downwarp forming the Great Valley of California, and beyond this are the Coast Ranges. In the south the latter are a series of northwest-trending fault slices almost parallel with the coastline, but in the north they become folded ranges. Also in the north are the Klamath Mountains, formed by upwarping. In southern California, associated with the Coast Ranges in a highly complex pattern of faulting, are the Transverse Ranges, including the San Gabriels and the San Bernardinos, and the Peninsular Ranges, which run down into Baja California. During the late Cenozoic uplift of these structures two large troughs developed nearby: the Great Valley and the Gulf of California.

California's block mountains are of relatively recent vintage, showing

The Sierra Nevada fault block: The block is about 400 miles long, and its western edge is at a depth of perhaps 5 miles beneath the surface of the Pacific Ocean.

Erosion panorama in the Sawtooth Range: These much-faulted mountains of southern Idaho rise to more than 12,000 feet. Sharp peaks and ridges are evidence of glaciation.

faceted scarps, hourglass valleys, and slickensided rock faces. To the north, in southern Oregon, a group of fault blocks raised even more recently are seen in the vicinity of the Warner, Abert, and Klamath lakes.

In northwestern Wyoming rise the snow-capped peaks of the Teton Range, sometimes called the American Alps because of their extreme steepness and deep glacial carving. They are the remnants of a fault block on which Precambrian crystallines are bared. Around 50 million years ago this block, about 60 miles long north to south and 20 miles wide, was tilted to make a scarp on the eastern side. Millions of years of exposure to the elements, with little uplift, cut the block low, but about 20 million years ago uplift was resumed and the terrain east of the scarp sank, forming Jackson Hole. During the new cycle of erosion uplift has continued, so that some of the peaks exceed 13,000 feet, rising more than a mile above Jackson Hole. The upper areas of the eastern faces of the mountains represent the original fault scarp; the lower areas are the dissected fault-line scarp resulting from the current erosion cycle. The instability of this region is evidenced by volcanic activity and earthquakes in the Yellowstone area just to the north.

Batholithic Mountains

Some of our most imposing highlands consist of uplifted masses of igneous rock. The largest of these have been formed at depth by the cooling of gigantic reservoirs of magma called batholiths (from Latin, "stone of the depths"). By definition batholiths have a diameter of forty miles or more. All these structures are granitic and are very old, dating back to the Mesozoic or earlier, but the uplifts have been more recent. Batholithic rock resembles the crystalline cores of folded mountains, and it is not always easy to distinguish between them. Some batholiths occur, however, where no other mountains exist.

As batholiths rise they cause an arching of overlying rock, which is then removed by accelerated erosion so that the batholith itself is exposed. The great fault block of the Sierra Nevada is essentially a batholith. Another is associated with the Zuni Uplift of northern New Mexico, which covers 1500 square miles and has a large granite exposure in the middle. Still larger is the Idaho Batholith, now exposed over an area of 16,000 square miles; it is the matrix for the Clearwater, Salmon, Sawtooth, and Coeur D'Alene mountains. But the greatest in North America is the Coast Range Batholith, which spans over a thousand miles from British Columbia to the Yukon Territory. This forms the Coast Range of Canada.

The nature of batholithic domes is well represented by the Black Hills of South Dakota. Here an elliptical mass of Precambrian granite was

Coastal mountains in southeastern Alaska: This extension of the great granite Coast Range of British Columbia reaches to over 6000 feet. Glaciation has been severe. The view is up the fiord of Tracy Arm.

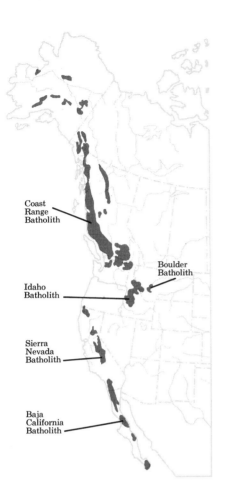

Batholiths of the Far West: Out of these uplifted masses have been carved some of the continent's most imposing mountain landscapes.

heaved up during the disturbances that raised the Rockies. The overlying strata were arched over an area measuring about 125 square miles north to south and 65 miles east to west. Originally the dome may have had an elevation of 10,000 to 11,000 feet, but the elements have cut away several thousand feet, and today, perhaps 50 million years after the uplift began, the highest elevation is 7242 feet. This is reached by Harney Peak, a granite spire rearing about 4000 feet above the neighboring lowlands.

The core of the Black Hills dome has been uncovered in the eastern sector over almost the full north-south span. Bordering the dome are the uptilted edges of sedimentary strata, which grow successively younger away from the dome. On the east, beyond a limestone ridge, is the red Spearfish shale valley and then the younger Dakota sandstone hogback. In the dome's western sector is a zone of schist overlapped by a wide limestone plateau; next come the red shale and the sandstone. Streams running down from the high, resistant central area are fed by tributaries

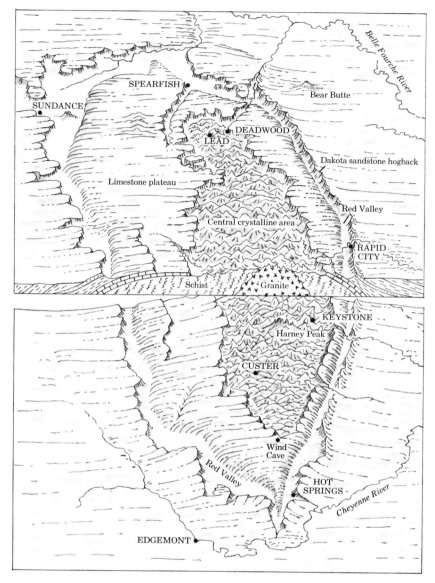

The Black Hills dome: The topography is strongly influenced by the difference in resistance between the core and the surrounding strata.

that follow weak belts in the sedimentaries ringing the dome. Several superposed streams reach the lowlands through water gaps cut into the Dakota sandstone ridges.

Like other batholithic domes the Black Hills structure is maturely dissected. Its uplift began, apparently, some 59 to 60 million years ago with the rise of the Rockies. River terraces and some parklike areas in the central region indicate three phases of uplift, separated by periods of stillstand during which erosion dominated.

The Eastern states have batholithic domes that in size and relief rank almost with the Black Hills. One is the Adirondack Dome of northern New York, spanning 7000 square miles. The central region of anorthosite, a feldspar-rich rock, rises more than 4000 feet above the lowlands, the highest elevation being Mount Marcy's 5344 feet. The anorthosite dates from Precambrian time and stands as highlands today because of its resistance and the intermittent uplifts. Forty-six of the peaks exceed 4000 feet. Fringing the dome are younger, uptilted metamorphics and sedimentaries.

Batholithic rocks of the late Paleozoic Era form crystalline highlands in central Connecticut, Rhode Island, New Hampshire, Maine, and Nova Scotia. The highest summits are in the White Mountains of New Hampshire, where Mount Washington rises 6280 feet. Quite as impressive is Maine's Mount Katahdin, a great isolated granite block with an elevation of 5268 feet. In the slopes of the White Mountains and Mount Katahdin are well-cut Pleistocene cirques and U-shaped valleys, and the felsenmeer on the bare summits testifies to the frost work here during recent time.

The south-central United States has its Ozark Dome, cut up into clusters of peaks of which the St. François Mountains are the highest, with a relief of 800 to 900 feet. The ancient igneous core of this dome is still mostly hidden.

The Adirondacks of New York: Relatively flowing profiles characterize these highlands carved from a dome of resistant, homogeneous, granitelike rock and later smoothed by Pleistocene glaciers. The view is eastward from Wright Peak toward Giant Mountain, noted for its great cirque and extensive slides.

An ancient range worn low: The Great Smokies, despite their billion years, include among their hazy summits the highest elevations in the eastern United States.

The White Mountains of New Hampshire: Sculptured by streams from a mass of uplifted igneous rock, these were strongly glaciated in the Pleistocene. The view is up the Great Gulf from the north side of Mount Washington.

Stocks and Laccolithic Mountains

Batholiths grade into the smaller frozen-magma reservoirs called stocks, which range up to forty miles in diameter. The Henry Mountains of southeastern Utah are exposed portions of stocks from which tongues of igneous rock reach out in many directions. Peaks carved from stocks are seen also in the Judith and Crazy mountains of Montana.

Upward-pushing basaltic magma has here and there raised surface strata to make a dome. Such domes have already been explained as folded structures. The igneous mass, called a laccolith (from Latin, "lake of stone"), is soon bared by erosion. Some laccolithic domes in the West are of mountainous bulk, with diameters of miles and heights of several thousand feet. They are encountered mostly in areas just east of the Rockies where the relics of Cenozoic volcanism have not yet been eroded away. Around the northern edge of the Black Hills are clusters of laccoliths—Crow Peak, Elkhorn Peak, Crook Mountain, and Little Sun Dance (Green) Mountain. The latter still carries a mantle of sedimentary strata.

Farther west is Devils Tower (Mato Tepee), a 1200-foot pillar called a laccolith by some geologists and a volcanic neck (the core of a volcano) by others. The rock is phonolite porphyry. The spectacular high, curving columns of hexagonal cross section formed by the jointing are among the finest of their kind.

The Boston Mountains of Arkansas: This section of the Ozarks is the exposed core of a very old range. A cliff on White Rock Mountain is seen here.

A laccolithic mountain in Utah: The sedimentary strata that once covered Navajo Mountain have long since been eroded away, leaving the igneous core exposed.

Scattered over the Colorado Plateau are other laccolithic domes, including in eastern Utah the Abajo Mountains, with three thousand to four thousand feet of relief, and in central Montana the Highwood Mountains. Among the latter are Square Butte and Round (Palisade) Butte, with a relief of about eight hundred feet. Another notable laccolithic peak is Navajo Mountain, in southern Utah near the junction of the Colorado and San Juan rivers. This one, like Little Sun Dance, retains its sedimentary cover.

Mountains of Circumerosion

Some peaks of mountainous size originated not by upfolding or upfaulting but simply by being cut out of broad, high-standing land masses. Although not true mountains they are often called mountains of circumerosion, or residual mountains, because they have been shaped by erosion.

Examples in the West include—among many—the Beartooth Range of Montana and the mountains of the Zion National Park area in Utah. In the East some notable examples are the Catskills of New York and the Poconos of Pennsylvania, both of which have been cut in the edge of the Allegheny Plateau. The steepness of the scarp, which rises several thousand feet, has enabled streams to cut impressive gorges into the sedimentary rocks. The dissection process has been accompanied by some uplift.

Likewise classed as mountains of circumerosion are the isolated hills known in the United States as monadnocks. A monadnock is an isolated peak on a terrain of generally low relief. It may survive either by virtue of superior resistance or because of remoteness from main drainage lines.

An Australian monadnock: Mount Olga, in Northern Territory, is a mass of strong conglomerate. The formation rises about 1500 feet above the plain.

Mount Monadnock in New Hampshire, from which such mountains get their name, rises some fifteen hundred feet above a plateau. It consists of andalusite schist, a particularly resistant metamorphic rock. Stone Mountain, Georgia's famous monadnock, is of granite, as are many other isolated peaks such as Pikes Peak in Colorado, the Brocken in the Harz Mountains of Germany, and the Cornwall hills in southern England.

Monadnocks appear often in clusters along the fringes of very old mountain ranges. These clusters are called unakas, after the Unaka Mountains of North Carolina. Linear monadnocks that rise above lowlands are known as catoctins, after Mount Catoctin, one of a chain of highlands along the Piedmont in Maryland and Virginia. The largest catoctin in this region is the Blue Ridge of Virginia, west of the Piedmont and east of the Great Valley.

Steep-sided, knoblike monadnocks that rise very abruptly from plains are known as inselbergs (from German, "island mountains"). They are typical of rain-forest areas of Southeast Africa. Prominent monadnocks on desert terrains, as in southern Arizona, also have been called inselbergs.

In the western United States some of the flat-topped highlands familiarly called mesas are analogous to monadnocks. Many consist of either resistant rock or weak rock capped by a resistant layer; but some survive because they happen to be far from major streams. Other highlands analogous to monadnocks are the residual masses of volcanic rock known as volcanic necks and lava buttes.

Extinct volcanoes in the Aleutians: Craters on Semisopochnoic Island have been active in very recent time. They form part of the volcanic "ring of fire" around the Pacific.

VII

Volcanic Scenery

Ringing the Pacific Ocean, forming fiery belts in southern Europe and Africa, and appearing spottily elsewhere on the globe are the zones of volcanism. Here are the weird landscapes of volcanoes, lava flows, fumaroles, hot springs, and geysers—scenery created in violence, suddenly, and subject to change overnight. These are areas of instability, of earthquakes and mountain building, where the crust is shifting and magma from miles-deep reservoirs is finding its way to the surface. Here new rock is being formed and the continents are being enlarged.

Our tour of mountains was largely land-based; mountains were considered as highlands on land. But if relief is measured from the ocean bottoms, it becomes apparent that mountains of tremendous stature rise all around the edges of the Pacific Basin and from its center, and also along other continental margins, especially in the West Indies, in the North Atlantic, and off the west coast of Africa. These mountains are volcanic. Some of them rise two or three miles above the ocean bottom and miles more above

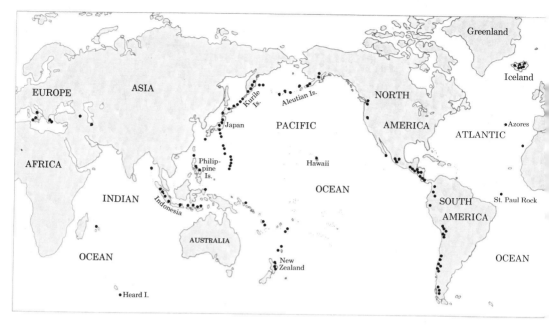

Distribution of the world's volcanoes: On a small map not all volcanoes can be shown independently, but their general distribution is apparent.

sea level. Hawaii's Mauna Loa, in the mid-Pacific, has a total relief of about thirty thousand feet and thus appears to top even Mount Everest. Clearly, volcanic land masses form important parts of the crust.

Approximately five hundred volcanoes have been active in the modern era, but a far greater number have erupted during the past few millions of years and are now dormant or extinct. All these are but successors to the uncounted, now-vanished volcanoes of the remoter past. We recall that the continents originated apparently by volcanism in the ocean bottoms and have since been enlarged by volcanic land-building outbursts along their margins. In the interiors the task of building is all but finished, and few traces of igneous activity remain. In the vicinity of young- to middle-aged mountain ranges nearer the coasts, such as the Rockies, volcanic cones and lava flows of fairly recent vintage are seen. There may be, as in Yellowstone, a few flickers of activity still. But it is mostly along the Pacific coasts and in the island arcs that volcanism persists. Here the continents are still in the making.

The relics of volcanism have in most regions vanished without visible trace. A few million years of erosion is usually enough to level volcanic highlands, especially in humid climates. The stumps of many old volcanic cones and the remnants of lava flows have been buried under rock wastes from nearby highlands. The roots of volcanoes dating from the early days of Earth lie deep in the continental cores. And so it is not easy for us to imagine the fiery dramas that were enacted long ago on perhaps the very ground where we stand today.

The Anatomy of Volcanoes

Volcanism, which is the eruption of lava through the crust, expresses disturbances in Earth's interior. The mantle is apparently being stirred constantly by convection currents, which carry heat from the planet's deep interior to the crust. This heat, supplemented by heat from radioactivity in the crust and perhaps also from crustal movements, melts masses of deep-lying rock. The melts mix with mantle material to form magma, and this, under the weight of miles of overlying rock, is squeezed upward through fractures in the crust. Some magma never reaches the surface, but mingles with rock picked up along the way and solidifies to form the intrusive igneous rock bodies known as plutons. Magma that reaches a zone just below the surface but does not break through accounts for landscapes of fumaroles and hot springs which, like Yellowstone, lack active volcanoes today. When magma does break through the surface—thus becoming lava—there is a volcanic eruption.

Lava erupting periodically from a pipelike channel may form a volcano. Such channels are found and enlarged by magma as it moves up through the crust. Often the vent of a volcano is at a point where fractures intersect. Where a chain of such points exists, a chain of volcanoes may be built up in time.

A volcano's form, as well as its usual mode of eruption, is determined primarily by the kind of lava it produces—basic, acidic, or andesitic. Basic lava, the kind that forms basaltic rock, contains less than 52 per cent silicon dioxide and is rich in ferromagnesian minerals. Being relatively fluid at temperatures characteristic of emerging lava (about 2000 degrees Fahrenheit), it flows from a vent with little explosive effect. Such flows, moving at five to ten miles per hour, and rarely as much as twenty-five miles per hour, may travel thirty to forty miles before cooling and solidification bring them to a halt. Accordingly, volcanoes that produce mainly basic lavas

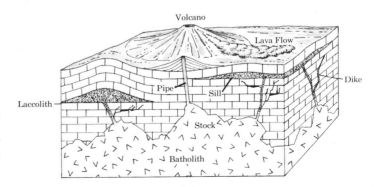

A volcanic region: Igneous intrusions and volcanic eruptions occur in many variations on these basic forms.

become domes or "shields" of wide diameter and gentle slopes. The Hawaiian volcanoes, models of this type, have slopes as gentle as 5 degrees toward the summits and not more than about 12 degrees at the base.

Acidic lavas are rich in silicon dioxide (over 66 per cent), aluminum oxide, and alkalies. They contain more gas than basic lava, are more viscous, and are quicker to solidify. These fast-hardening lavas tend to clog the vent, and periodic explosions occur as the volcano tries, as it seems, to clear its throat. Much of the lava is blown to fragments—some into dust that rises miles high and is carried hundreds or thousands of miles by winds, and some into heavier material that falls near the crater. These fragments, called pyroclastics, become mixed with flows of lava that harden before they travel far down the slopes. Thus the volcano's output accumulates mostly near the vent, making a steep-sided cone with slopes typically as much as 40 degrees at the summit (where the largest fragments fall) and nearly 30 to 35 degrees farther down, where the material is finer. Such are the cinder-cone volcanoes, common especially in the Mediterranean region and the Americas.

Most lavas consist of andesite, which contains 53 to 65 per cent silicon dioxide and is intermediate between basic and acidic lavas. Andesite is erupted by volcanoes along a line running north from the Andes of South America, through Central America and the western United States, on into Alaska and the Aleutians, and down the western side of the Pacific through the islands of Japan, the Philippines, New Guinea, and New Zealand. Nearly all these volcanoes display cones with much steeper sides than are

The perfect cone: Erupted materials have built Japan's Mount Fuji to 12,395 feet.

Cinder cones at Craters of the Moon in Idaho: These were formed by lava froth spewed out about 2000 years ago. The region is now inactive except for a few hot springs.

seen on the Hawaiian domes. Japan's lovely Fujiyama, last active in 1707, is the type.

Volcanoes generally pass through phases in which basic, andesitic, and acidic lavas are erupted alternately. During explosive periods large amounts of pyroclastics accumulate on the slopes. But explosions weaken the cone and may open fissures in its sides, and through these openings lava escapes. Flowing over the pyroclastics, it binds them into solid masses, and so a composite cone, called a strato-volcano, is gradually developed. Most large volcanoes—those with a long history—are composite; the main exceptions are those of Hawaii.

Subsidiary vents usually develop on a volcano's flanks. Some are pit craters formed where small areas of the cone collapse because of the removal of material from inside by eruptions. Spatter cones are built up by gassy outbursts and resulting spatter around small vents. Gas "leaks" through narrow cracks in the cone become fumaroles. The main crater, which may be only a few yards wide at the volcano's birth, can grow in a few weeks or months to a diameter of hundreds of feet, and it may be enlarged even more by the wear and tear of continuing eruptions.

Prolonged eruptions may undermine the crater and cause collapses of its walls and floor. The crater thus may evolve into the broad, deep basin known as a caldera, or volcanic sink, which is sometimes miles wide. From its floor small domes called tumuli may be raised by pressure from below. Where new vents open in the caldera and on the mountain's slopes, new cones may build up, some forming nested patterns.

The lavas of all volcanoes, except in the Pacific Basin, become more acid with time. Originally a volcano's stock, or magma reservoir, is basic in composition, like the mantle below. But the stock engulfs more and more granitic and carbonate material from the crust, and this mingles with the basic lava; thus the andesitic lavas are formed. The Hawaiian volcanoes keep producing basic lavas simply because granitic and carbonate materials are scarce in the crust around them.

As long as the magma keeps coming up, a volcano remains active and its growth can continue. Over the centuries some volcanoes attain heights of miles. However, they usually become more explosive and self-destructive with age. Finally, battered after millennia of activity, they quiet down for good. The elements then take over and erase the once-fiery peaks from the face of the land.

Volcanic Ejecta

Magma is a hypothetical substance; no one has ever seen it. What the human eye sees is a material which, on being relieved of pressure as it arrived at the vent, and on reacting chemically with the air, was transformed into lava. In the process of cooling, the material loses most of its gas content. On hardening, it forms rock that consists of the familiar minerals of Earth's surface.

In acidic lavas, the gases usually exceed the solids by weight. Great amounts of steam are mixed with carbon dioxide, nitrogen, and gaseous sulfur compounds, along with some hydrogen, chlorine, and carbon monoxide. Basic lavas also contain these gases, but in smaller quantities.

Pyroclastics vary widely in size and shape. The largest fragments, called volcanic blocks, are rock chunks from the pipe or the crater. Gobs of lava that are hurled through the air and cool in spindle- or pear-shaped forms are known as volcanic bombs. Smaller gobs, up to the size of a walnut, are lapilli (Italian, "little stones"), Pelé's tears (after the goddess of the Hawaiian volcanoes), or Apache tears. Pea-sized fragments are cinders, and highly pulverized bits are called volcanic ash or dust.

Some acidic lavas while still liquid are made very frothy by gases escaping from them or being blown through them. Such lavas cool to make the very light, porous rock called pumice—the familiar household abrasive. It was pumice, which is produced in great volumes by highly explosive eruptions, that buried Pompeii in A.D. 79.

Lava that has been blown into stringy forms has been called Pelé's hair. Rock fragments bound together by cooling liquid lava are known as volcanic breccia.

On the slopes of a volcano and over the countryside for miles around, the small-caliber pyroclastics—cinders, ash, dust—may accumulate in thick beds. Often these become consolidated by volcanic heat or percolating mineral waters to form the rock called tuff. This material lies deep around many explosive volcanoes, such as Mount Mazama in Oregon and the volcanoes of northern New Mexico and Arizona.

An eerie feature of some volcanoes is the plug dome, or spine, consisting of a towerlike mass of very viscous lava that is slowly pushed up out of the crater. Some domes attain heights of hundreds of feet. The additional weight on the cone, which is already subject to the pressure of gases inside, may cause a side to blow out.

Flows of acidic lava form the light-colored rock rhyolite, similar chemically to granite. If the cooling is too rapid for crystals to form, the rock called obsidian, or natural glass, results. Extremely gassy flows may form pumice.

Andesite, the rock intermediate between the acidic and basic types, is greenish-gray. Often it displays sizable crystals of feldspar and darker minerals. These mingled crystals are known as phenocrysts, and the rock they form is called porphyry.

Ropy pahoehoe: These rock coils at Craters of the Moon formed from nongassy lava.

A lava cave at Lava Beds National Monument, California: Lava caves, tunnels, and tubes may be formed by crusts that develop on flows of pahoehoe.

Flows of basic lava show a diversity of forms. A very liquid lava may cool in smooth, thick, glistening coils, called pahoehoe (puh-HOY-hoy) in Hawaii. When this lava flows into water and cools there, it may assume the spheroidal shapes appropriately called pillow lava.

The surface of a gassy basaltic flow often cools as a porous, cindery material called scoria. If this open-textured rock lies buried for a long time, ground water may fill the pores, leaving glittering crystals of calcite, zeolite, quartz, and other minerals.

The surface of a relatively viscous basaltic flow solidifies much faster than the lava inside. The result is a thin, hard, glasslike crust which becomes broken and jumbled on the flow. To this jagged surface Hawaiians have given the name aa (AH-ah). Sometimes a flow breaks out of a large shell formed by rapid surface cooling, and the shell is left as a lava tunnel or lava cave, complete with lava "stalactites" and "stalagmites" formed by drip from the ceiling.

Certain basalt flows develop columnar jointing as they contract upon cooling. The columns are perpendicular to the cooling surface and their ends appear at the surface as hexagonal or pentagonal patterns. Columnar jointing is especially remarkable at Giant's Causeway in Northern Ireland and Fingal's Cave in Scotland. The United States has good displays in the Northwest and the Southwest, at Devils Tower in Wyoming and Devils Postpile in California. On the European continent there are fine columnar basalts in Saxony, Germany, and in the Auvergne of France.

Erosion on Volcanoes

The profile of an active volcano is determined mainly by its eruptive habits. During intervals of inactivity, however, a radial pattern of drainage develops on the slopes. This represents the most efficient drainage on materials that are relatively homogeneous. The gullies and ravines are deep and narrow, in accordance with the steepness of the slopes and the weakness of the rocks. If the material of the slopes is poorly consolidated—and volcanic ejecta usually are—rainwater will work down into them easily, perhaps emerging at lower levels as springs.

Accumulations of lava dust on the cone are gullied and washed downslope rapidly. When an eruption produces large volumes of steam, condensation causes torrential rains, which turn the dust into mud and sweep this

Giant's Causeway: This rock city of basalt lava columns, on the northeastern coast of Northern Ireland, is one of several in British Isles. Most columns have six sides.

down the mountainsides in overwhelming quantities. It was a mudflow that buried the town of Herculaneum during the eruption of Mount Vesuvius in A.D. 79.

In cold climates, glaciers may develop in valleys high on a volcano's slopes. The descending ice digs into the sides of the mountain deeply because of the vulnerability of the rocks. During eruptions much of the ice melts, and the water is added to condensed steam to produce floods and mudflows.

During periods of activity, the drainage system on the slopes of a cone may be deranged by lava flows, showers of pyroclastics, and the formation of fissures and sinks. After each eruption, drainage adjusts to the new contours. Eventually, as the volcano becomes extinct, drainage becomes well integrated. Erosion then dissects the cone quickly and soon levels it. In a dry climate, such as that of the American Southwest, cones may still be intact after a few thousand years, but elsewhere they are less resistant.

Volcanoes in North America

Most North Americans must travel a long way to see a live volcano. East of the Mississippi a few scattered hot springs, such as those at Hot Springs, Georgia, are the only active evidences of Earth's internal heat. West of the Mississippi one begins to encounter volcanic relics only near the eastern edge of the Rockies, from Montana to New Mexico. Among the Rockies proper there are clusters of sizable extinct volcanic peaks, such as the Crazy Mountains of southwestern Montana, the San Juans of southern Colorado, and the San Francisco Peaks of northern Arizona. Wyoming's Yellowstone region is but slightly active today, and the Craters of the Moon locality in Idaho, where cinder cones were active until five hundred years ago, is quiet. Capulin Mountain in northeastern New Mexico, active two thousand years ago, and Sunset Crater in northern Arizona, active until about A.D. 1060, are cinder cones in a fine state of preservation, but extinct. No volcano classified as active is encountered until one is almost at the Pacific Coast. There, in northern California, is Mount Lassen, which was active as recently as 1921.

Lassen, 10,457 feet high, is the southernmost large peak of the Cascade Range, which is by far the greatest volcanic chain in the continental United States south of Canada. At least 120 volcanoes are numbered in this chain, which stretches from northern California about five hundred miles into southern British Columbia. Originating in the middle Cenozoic, the Cascades were built up rapidly and some were active into the Pleistocene. Today they are old as volcanoes go; perhaps their fires will not revive. But as mountains they are young, with still-lofty summits such as those of Mount Rainier, 14,415 feet high, and its rivals, Mounts Baker, Hood, and Shasta.

South of the Cascades minor volcanic features are seen particularly around Death Valley. In southwestern Mexico an active volcanic chain starts west of the Sierra Madre Occidental and continues on through Central America into South America, which has the highest volcanic peaks in the hemisphere. North of the Cascades in British Columbia one enters a zone of quiet, but in Alaska there are many youthful volcanoes. A long chain, beginning with Mount Katmai at the eastern end of the Alaskan peninsula, includes thirty-six recently active peaks in the Aleutians, from which the chain runs on down the western side of the Pacific.

Eastern North America has seen no volcanism for two hundred million years. The relics of very ancient volcanism here have been destroyed by erosion or buried by sediments of recent eras. Among the Appalachians and other eastern highlands, erosion has had time to sweep away most of the old Triassic lavas. But on certain lowlands, notably those of Massachusetts, Connecticut, New Jersey, and Maryland, some of the flows were buried long ago by erosional debris washed down from the mountains. After tens of millions of years the old lavas have been uncovered again. Typical of the ridges and mesalike hills formed where these flows cap the weak red Triassic sandstones and shales are the Watchung Hills of New Jersey, several similar highlands in central Connecticut, and the Mount Tom and Holyoke ranges in Massachusetts.

In the Cascade Range: North *(left)* and Middle Sister peaks are typical of the great volcanic mountains of the Northwest. In the foreground is a field of scoria.

Profile of a shield volcano: Mauna Loa owes its breadth and mostly gentle slopes to the fluidity of its lavas. Its shield at sea level has an area of about 2000 square miles.

Madame Pelé's Mountain

Hawaii's Mauna Loa is the greatest of all volcanoes, at least in sheer bulk. Its volume of 10,000 cubic miles makes it also probably the world's largest mountain.

The rise of Mauna Loa started perhaps ten million years ago with flows of basalt from a long rift zone in the Pacific Basin—the same zone that has produced the other volcanoes of the 1500-mile-long Hawaiian archipelago. Intermittent eruptions since then have built the mountain 15,000 feet up to sea level and 13,680 feet above it. In the summit is a great caldera some 3 miles long and 1½ miles wide, with a maximum depth of 600 feet. This is the original crater as enlarged by explosions, collapse of the crater walls, and subsidence. On the mountain's flanks are pit craters, due to subsidence, and spatter cones. The very liquid flows of the past have formed a dome at least 60 miles long and about 30 miles wide. In the eruption of 1859 one flow traveled 35 miles before plunging into the sea.

Eruptions today are featured by lava jets rising as much as 1000 feet above the crater and by fast-moving flows, usually from fissures in the crater's sides. Some of the flows reach speeds of about 25 miles per hour near the source, but the usual speeds are much slower. Eruptions are preceded by earth tremors and a definite swelling or doming up of the mountain, but after large volumes of lava have been extruded—at rates up to five million tons per hour in the early part of the eruptive phase—the mountain gradually subsides again. Eruptions occur usually at intervals of three or four years, lasting a few days to a few weeks.

On the southeast slope of Mauna Loa is another volcano, Kilauea, built up by separate eruptions. Although much smaller it is an impressive mountain in its own right, with a total reach of nearly 20,000 feet above the sea bottom. In its caldera, which measures about 2 by 2½ miles, is the Halemaumau, or fire pit, the site of the famous lava lake which has accumulated over the vent at times in the past. The pit is the home of Madame Pelé, the volcano goddess.

Kilauea and Mauna Loa are the only Hawaiian volcanoes that have been active in modern times, but the islands of the archipelago are all volcanic. The islands northwest of Hawaii are older and have been cut down by erosion. Some, such as Midway and Ocean Island, which were long ago reduced to sea level, became capped with limestone produced by corals, as often happens in warmer seas. The relative lowering of sea level since then has left the caps exposed as coral islands.

Strabo's Scorched Peak

Mount Vesuvius, looming over the Bay of Naples in Italy, typifies the explosive volcano. Relatively young, it dates back about 10,000 years. Eruptions from a fissure in what was once a much larger Bay of Naples raised the cone to mountainous proportions within a few thousand years. Appar-

Mount Vesuvius and Naples: The volcano probably originated during the Pleistocene. The disastrous eruption of A.D. 79 which buried Pompeii followed centuries of dormancy.

ently the volcano was dormant for about 700 years before the Christian era.

The significance of the earthquakes that began in the vicinity of Vesuvius in A.D. 63 was missed until, in A.D. 79, the side of the mountain blew out. Within a few hours Pompeii, four miles from the summit, was buried by fifteen to twenty-five feet of pumice and dust. The volcanic showers were so hot and so laden with lethal gases that more than twenty-one thousand people were burned to death or asphyxiated as they cowered in their homes or fled for the waterfront. Neighboring Herculaneum, three miles from the summit, was buried a day or two later to depths as great as sixty-five feet by a volcanic mudflow, but only after most of the townspeople had escaped. The steaming lake of mud that covered the town eventually solidified to form volcanic tuff.

Herculaneum was thus made highly resistant to the picks and shovels of excavators. Pompeii made easier digging, and its mantle of ash and pumice was soon trenched by returning homeowners, looters, and farmers seeking building stones. Then the city was forgotten until stumbled upon and identified by antique-hunters in the eighteenth century.

After A.D. 79 Vesuvius erupted every century or two for nearly a thousand years. Then, after almost five centuries of sleep, came the great eruption of 1631, which flooded six towns with lava and nine with mudflows, dropped a foot of ash and cinders on Naples, and killed about four thousand people. Since 1631 the mountain has erupted with varying degrees of violence, though not catastrophically, at intervals of about seven years.

Since 1845, when the Vesuvian Volcanic Observatory was established high on the mountain, much has been learned about Vesuvius and volcanoes generally. Studies indicate that the top of the stock, or magma reservoir, is about fifteen thousand feet below sea level, or nearly four miles beneath the 3858-foot summit. The depth of the stock is believed to be ten to fifteen miles. Originally the magma mass lay much deeper, but the chamber has been enlarged upward by engulfment of rock from its ceiling. The absorption of this rock, which consists of limestone, shale, and sandstone, accounts for the acidic trend of the volcano during the past two thousand or more years.

Volcanoes That Destroyed Themselves

Although eruptions may blast away sizable portions of the cone, an active volcano usually erupts material faster than erosion destroys and scatters it. But volcanoes gradually degenerate as their lava becomes more acidic. Eruptions become more explosive and destructive, and eventually an outburst of extra violence may shatter the mountain.

Krakatoa in Sunda Strait, between Java and Sumatra, was once an island some five miles in diameter. During the past century it has been reduced by explosions to three little islands. These are the edges of the caldera, most of which is now under water. The outburst of 1883 scattered an estimated five cubic miles of pumice into the air. It fell on ships 1600 miles away, darkened the sky at a distance of 275 miles, and was carried round the world by the winds of the upper atmosphere, causing unusually red sunsets for two years. The sound was heard nearly 3000 miles away—as far as from New York to London.

Equally catastrophic may have been the eruptions that ripped off the top of Jemez Volcano, in north-central New Mexico, during the early Pleistocene. The remains of this one-time giant form today's Jemez Mountains. On their western side is the Valles Caldera, more than thirteen miles in diameter, possibly the largest such depression in the world. The masses of rhyolite pyroclastics blown out of the original crater, burying the countryside for miles around, have been estimated at fifty cubic miles.

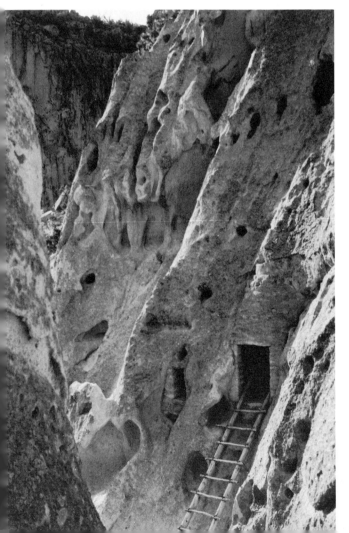

A canyon in volcanic tuff: Indians in Frijoles Canyon, New Mexico, easily cut cliff dwellings into this porous rock. The site is in Bandelier National Monument.

It was long thought that a tremendous explosion about 6500 years ago must have been responsible for the formation of the great caldera of Mount Mazama, in southern Oregon, which now holds Crater Lake. Present knowledge indicates subsidence rather than an explosion, but it was the loss of enormous volumes of material that produced the subsidence. Judging from the size of the truncated valleys on its flanks today, Mazama was a 12,000-footer, rising high in the volcanic chain which includes the still intact Mounts Rainier, Adams, Hood, and Shasta. The depth of the pumice scattered over a radius of thirty-five miles indicates that no less than ten cubic miles of it was erupted, and that the mountain lost as much as two miles of its height. Crater Lake has a diameter of about six miles and a maximum depth of 2200 feet. Around the lake the walls of the caldera rise 500 to 2000 feet. Wizard Island, near the edge of the lake, a cinder cone reaching 763 feet above the water, was built up by small-scale volcanism before Mount Mazama lapsed into complete inactivity.

Volcanic Islands

Most of the world's volcanoes are encountered along the island arcs that rim the Pacific Ocean. Some emerge as chains along the Mid-Atlantic Ridge and in the West Indies. These archipelagoes, along with a few isolated islands elsewhere, were built up—as were the Hawaiian Islands—by eruptions from the ocean floor. The volcanoes active in the Pacific ring today are generally of an explosive, andesite-producing variety, as are the volcanoes of the West Indies. Those of the Mid-Atlantic Ridge produce basaltic lavas. Most peaks along the arcs, including nearly all those in the West Indies, are extinct or dormant, but some have put on spectacular and disastrous exhibitions within our time.

The worst of these outbursts occurred in May, 1902, on Martinique in the West Indies. The side of Mount Pelée blew out, and a roaring cloud of incandescent gas and dust—a nuée ardente—rushed five miles downslope at a hundred miles an hour to engulf the city of St. Pierre. Of the town's population of nearly thirty thousand, only two—who were in an underground jail—survived. Later that year, after a period of quiet, internal pressures forced up through the crater the Tower of Pelée, an enormous plug or spine of viscous lava, 1020 feet high and as much as 500 feet in diameter.

Volcanic activity along island arcs within recent centuries has produced a number of new land masses. In the Aleutians the island of Bogoslof, first noticed in 1768, has been all but destroyed several times by explosions and by wave action, only to be rebuilt by new eruptions as a new island or island group. At last report it was a single land mass several miles long, with a steaming, mile-wide lagoon in the middle.

In October, 1957, a spitting volcanic cone rose up out of the sea next to

Fayal, an island off the Azores. After a month of relative quiet, during which its top was cut off by wave erosion, it was rebuilt by renewed activity to form a mile-long extension of Fayal.

In September, 1952, a Japanese fishing boat reported an active volcano rising out of the sea two hundred miles southeast of the island of Honshu. Two small government ships, each bearing scientific observers, headed for the scene. The first to arrive witnessed from a safe distance a number of tremendous explosions which destroyed the visible part of the cone. The other ship, arriving two days later, was exactly over the drowned vent when another great explosion occurred. The ship and all aboard were lost. For about a year the activities of this volcano, Myozinsyo, continued, but the island itself did not reappear.

Many volcanic islands are not recognizable as such, because they are covered with coral. Their activity ceased long ago, and their cones have been obliterated by erosion. But some volcanic island arcs may be still only near the beginnings of their careers. They may become the future edges of continents.

Extinct Volcanoes

As volcanoes age, they show it. Explosions may blow off the top of the original neat cone. Crater walls collapse, and the crater becomes a caldera.

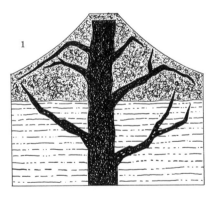
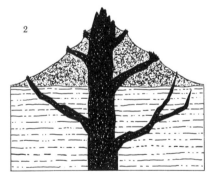

Erosional history of a volcanic neck: In the first stage the volcanic cone has already been much reduced by erosion. Later the dikes, being relatively stronger than the fragmented lavas that make up most of the cone, form ridges.

Parasitic cones grow up from fissures on the volcano's flanks, and new cones may form in the caldera itself. Erosion cuts deep ravines into the mountain's slopes. In cold climates, as in Alaska today and in the Rockies ten thousand years ago, the ravines are ground and gouged by glaciers.

Volcanic islands disintegrate swiftly under the attack of ocean waves and currents. In zones of active diastrophism, some islands break up and founder. Many, as in the Hawaiian Archipelago, survive merely as small, isolated stacks barely rising above the waves, or as submarine banks rapidly accumulating coral.

Extinct volcanoes everywhere erode so fast that they cannot survive for long in recognizable condition. Volcanic remnants visible around us today can only represent volcanoes of recent vintage, or ones that were buried long ago by sediments and but recently exhumed.

After a volcanic cone has been mostly eroded away, its core may still survive. This core, called a volcanic neck, consists of lava that cooled as a dense solid in the volcano's throat as it became inactive. Destruction of the weak cone material leaves the neck standing as an isolated, ragged rock tower, averaging one thousand to two thousand feet in diameter. Often one may see radiating from the neck a number of resistant dikes, formed originally as fissure fillings in the cone. Where activity was relatively recent, as in the Mount Taylor district of New Mexico and parts of the Colorado Plateaus, the necks rise boldly above the surrounding weak sedimentary rocks.

Volcanic necks in Europe have been favored through history as sites for fortresses, churches, and other special buildings. Edinburgh Castle in Scotland was built on a neck, as were the Chapel of St. Michel d'Aiguilhe and others in the old volcanic Auvergne area in central France. A well-known volcanic neck in England is Castle Head, in the Lake District.

Where volcanism ceased in the very remote past, as in the eastern United States, the volcanic necks have long since vanished. Here and there, however, bedded lavas and consolidated pyroclastics are exposed in the walls of deep valleys, on mountainsides, and in road cuts. Pillow lavas of Triassic age appear in road cuts and quarries near Paterson, New Jersey, and two small basalt hills, called volcanic necks by some observers and flow remnants by others, rise up from the soft clays and sandstones of the Hackensack Meadows a dozen miles to the east. Similar remnants of Triassic volcanism, some 200 million years old, are scattered from Massachusetts to Georgia.

Some vanished volcanoes have left ring dikes, such as those of the White Mountains of New Hampshire. A ring dike is an outward-dipping, collarlike wall of lava that filled a circular fissure opened around the edge of a caldera by collapse of the floor. These structures often are seen in concentric series, but always are incomplete—they do not form complete

A roadside volcanic neck in northwestern Colorado: This eroded volcanic feature, near Yampa, is typical of many in a region that now has no active volcanoes.

circles. Extending far below the original base of the volcano, they are now exposed long after the complete removal of the cone.

Fissure Flows

Lava erupts often through fissures in fault zones, and fluid lavas pouring from the fissures may spread sheetlike over very wide areas. Where the lavas are viscous, parts of a fissure become blocked, eruptions become more explosive, and cones develop. Chains of such cones cross Nicaragua, Iceland, and central France.

The Columbia and Snake River plateaus, which occupy about 175,000 square miles of the states of Washington, Idaho, Oregon, Montana, and Nevada, were formed by basalt flowing apparently from fissure networks.

Plateau-forming basalts: Successive layers near Palouse Falls, Washington, represent intermittent flows of the kind that built up the Columbia Plateaus.

These eruptions, which occurred 20 to 25 million years ago, overwhelmed the entire region to depths of hundreds of feet and, at the deepest, nearly covered mountains rising 4000 feet above the original plain. The Deccan Plateau of India and the Paraná region of South America were the scenes of even vaster basaltic flows. Ireland, northern Scotland, and Iceland were built up by lava outbursts from fissures in the bottom of the North Atlantic.

Massive fissure flows have occurred during historic times only in Iceland. This island is a hot spot on the Mid-Atlantic Ridge and Rift, which is part of the system of great cracks that divide the planet's crust. The vicinity of the ridge is often shaken by earthquakes, and undersea volcanism along this line has built up not only Iceland but many land masses to the south, among them parts of the Scotland and Ireland, the Canaries, and the

Azores. Iceland's volcanoes have been intermittently active, but the prime cause of worry has been fissure flows. The worst occurred in 1783, when fast-spreading basalt from a 10-mile-long cleft filled Skafta Valley over a distance of 50 miles to a width of 12 to 15 miles and an average depth of 100 feet, the maximum being 600 feet. More than half of the island's livestock perished, and the destruction of crops caused a famine. The human cost was one fifth of the population.

Fissure flows sometimes cover lowland areas to form lava plains, such as those of the Southwest. As streams cut through such a plain, the lava layer may form a protective cover over underlying weak sedimentary rocks, and thus a landscape of lava-capped mesas may evolve. Remnants of ancient lava plains in the West include the well-known Mesa de Maya and Raton Mesa east of the Colorado Rockies.

Fumaroles, Hot Springs, and Geysers

Fumaroles, hot springs, and geysers are features especially of dying volcanic areas. All result from the presence of magma near the surface.

Fumaroles are steaming fissures that appear in volcanic craters, on the sides of a cone, or on any other volcanic terrain. The vapors issue at very high temperatures, often around 1200 degrees Fahrenheit; they are about 99 per cent steam, mixed with a little carbon dioxide and still smaller amounts of hydrochloric acid, hydrogen sulfide, nitrogen, hydrofluoric acid,

Minerva Terraces at Mammoth Hot Springs: The rock of this formation in Yellowstone National Park is travertine, deposited by hot ground water from the depths.

oxygen, and ammonia. Some of the vapors derive directly from the magma, but much of the steam is from ground water vaporized by contact with magma or hot rock. As the vapors escape, minerals that have been held in solution by the great heat—silica, sulfides, and carbonates, for example—are deposited around the fissure.

Fumaroles are abundant in Yellowstone National Park, Wyoming. Mount Etna, in Italy, has about five hundred, many of which have the interesting habit of puffing simultaneously. The greatest display of fumaroles in modern times was seen in the appropriately named Valley of Ten Thousand Smokes, in Alaska, following Katmai's tremendous outburst in 1912.

Hot springs are pools consisting largely of ground water that has worked down to the hot zone and then risen by convection to the surface again. While it is at depth, some juvenile water (water originating directly from magma) may become mixed with it. On emerging at the surface these waters are normally well below the boiling point.

In Yellowstone the waters of Mammoth Hot Spring approximate about 170°F. The water of some springs in the park is believed to rise from depths as great as eight thousand feet. In southern Idaho, the springs are much hotter and may consist mainly of juvenile water.

Around hot springs are deposits of minerals dissolved by the hot waters underground and then precipitated by cooling at the surface. Springs that are full of this mineral matter, kept churning and bubbling by escaping steam, are called mudpots. Like fumaroles, hot springs are common even in zones where large-scale volcanic phenomena are missing—for example, Hot Springs in Arkansas and The Geysers in California.

In a very few localities, notably Yellowstone, New Zealand, and Iceland, special conditions produce geysers. These are essentially erupting hot springs. Beneath each geyser is a long channel with loops and side chambers that act as traps for ground water. As water in the main channel becomes heated and its density decreases, it tends to rise, whereupon

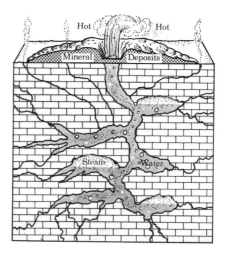

Anatomy of a geyser: Superheating of ground water in "traps" causes periodic eruptions.

denser, cooler water comes down the channel to replace it. But the traps hinder circulation. Water in the lower traps becomes extremely hot, yet because of the pressure of overlying water its change to steam is delayed. Finally some of this water does get hot enough to flash into steam; the steam drives water up through the channel and out the top; the relief of pressure on the near-boiling water lower in the channel allows this also to turn to steam; and so, in an accelerating chain reaction, a large volume of the water in the channel quickly turns to steam and erupts violently into the upper world.

Waters ejected by geysers are very near the boiling point. While underground their temperature enables them to dissolve sizable amounts of minerals. These are precipitated around the vents, building up various odd forms to which tourists and park officials give fanciful names. The Yellowstone landscape is covered with steam- and water-deposited minerals. Some of them are lethal to vegetation and have destroyed extensive woodlands.

Magma That Cooled Below

Some masses of magma cool within the crust without reaching the surface. As the "plutonic" rock thus formed is uncovered by erosion, it yields some remarkable landscape features. The greatest of these are the batholiths, stocks, and laccoliths, which form mountain ranges and lesser highlands. Some acquaintance with these was gained in our tour of mountains. But other, smaller plutonic forms which account for minor topographical features demand recognition here.

A sill forms where magma penetrates between strata of the existing "country rock," cooling there as a slab like the slice of cheese in a sandwich. When exposed by erosion or faulting, the sill influences topography because it is usually more or less resistant than the strata above and below it.

Sills when exposed by erosion may become the caprocks of highlands. Such highlands, often seen in the form of mesas in the West, occur as hogback ridges or cuestalike slopes along the Triassic lowlands of the East, from Massachusetts to Georgia. The gentler slope of the ridge follows the dip of the sill, and the edge of the sill outcrops at the top of the steeper slope. In the East the profiles tend to be rounded in the humid climate and are usually mantled by vegetation, so that the caprock may not be readily noticed.

A thick Triassic sill is responsible for the scenic Palisades of the Hudson River, a 20-mile-long cliff of diabase (a basaltic rock) on the west shore of the river northwest of New York City. This enormous slab, dipping gently westward and reaching a thickness of 900 feet, protects the weak underlying sandstones, which are locally exposed just above the river's level. The cliff is being cut back by erosion, but because the red sedimen-

The edge of a sill: The Hudson Palisades as seen southeastward from High Tor, which rises nearly 800 feet above the river. In the foreground the diabase rock of the sill is exposed in pentagonal and hexagonal patterns, which are the ends of columns. A quarry is visible in the middle background.

taries are weaker than the overlying diabase, the steepness of the cliff is maintained.

A prominent feature of northern England is the Great Whin (basalt) Sill. This structure, with interesting basalt columns, is exposed at many points between northern Northumberland and the Farne Islands. From the latter it runs crookedly southwestward, and then turns southeastward along the west side of the Pennine Hills.

Another kind of intrusion between strata is the phacolith. This is confined to the crest of an anticline (upfold) or the trough of a syncline (downfold). As the fold is cut away by erosion, the difference in resistance between the phacolith and the country rock of the fold is expressed as either a ridge or a valley.

The lopolith also is an intrusive feature. It is a basinlike structure in which the central depression is due to subsidence caused by withdrawal of supporting magma.

A magma mass that cuts through the rock strata instead of squeezing

stle Geyser in Yellowstone: The minerals deposited
und the vent are mainly silica.

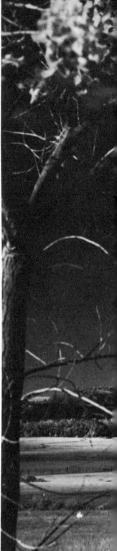

A dike in the Hudson Highlands: This basaltic intrusion appears as a black band running diagonally across the structure of the gneiss. It is being weathered somewhat faster than the host rock.

between them makes a dike, which is characteristically a wall-like form that may be inches or scores of feet in thickness. It may be either horizontal or inclined, according to the structure of the country rock. As dikes in relatively weak country rock are uncovered, they form ridges, as on the Colorado Plateaus. Many such structures are seen in the vicinity of the San Juan Mountains of Colorado. Where the country rock is stronger than the dikes,

Devils Tower in Wyoming: The 865-foot tower is probably a laccolith but could be a volcanic neck. According to Indian lore, the "grooves" were produced by the claws of a giant bear trying to seize an Indian maiden who was on the top.

these are often cut away by streams to form ravines or gorges, as in the granitic highlands of the Northeastern states. The Flume at Franconia Notch in the White Mountains of New Hampshire is the site of one such destroyed dike. Equally well known in the Adirondack region of New York is the great eroded-out dike in Mount Colden.

An alluvial plain with volcanic features: The Great Plains area of New Mexico is dotted with extinct cones and patched with lava flows. This photograph was taken southwestward from Capulin Mountain.

VIII

Plateaus and Plains

Traveling over a terrain such as the Great Plains, after leaving mountain country, one wonders that such level places exist. But if they appear never to have suffered deformation, volcanism, or severe erosion, they are deceptive. They are flat not by original creation but as the result of the same land-shaping processes that go on elsewhere.

Earlier it was said that some flatlands are flat because they are underlain by horizontal rock strata. But many have level surfaces for other reasons. Flood plains and deltas, such as those of the lower Mississippi, are level because of the wide, even deposition of sediments by streams. The Columbia Plateaus of the Northwest were formed by successive highly fluid lava flows. The extremely level character of some plains in Iowa, Illinois, and Indiana is due to the even spreading of glacial debris by floods of meltwater. The Great Plains, which have nearly horizontal strata beneath their soils, were further smoothed by the eastward advance of vast alluvial fans from the Rockies. But peneplains, as explained earlier in the discussion of rivers, have been worn flat by erosion.

The retreating edge of a plateau: Giant monoliths have been cut from the margin of this Colorado highland by swift streams, which probably have followed joints.

Neither plains nor plateaus are uninterruptedly flat. Plateaus, though often showing a horizontality of structure, have been more or less cut up by streams because of their high elevation. Coastal plains often have seaward-facing cuestas, such as those of New Jersey and the Gulf Coastal Plain. Hogbacks as well as cuestas are seen on plains at the edges of domes and mountains—around the Black Hills, for example, and facing the Front Range in Colorado. Small, resistant basalt hills dot plains in Connecticut, Massachusetts, and New Jersey, and volcanic necks rise from the Colorado Plateau. Where a cuesta is retreating before erosion, as on the Great Plains, it may leave outliers, or isolated rock masses, like rear guards behind it. Glaciated plains may be interrupted by rock-basin and kettle lakes, and ridges of sand and gravel deposited by melting ice.

Plateaus: High Flatlands

The term "plateau" is usually applied to high, extensive, flat terrains of nearly horizontal strata associated structurally with adjacent mountain masses. The high elevation of such a terrain is due to uplift. The plateau is cut by deep valleys, because as uplift occurs the streams are rejuvenated, and by the time much elevation has been gained, dissection is well advanced. So-called lava plateaus are terrains built high by lava flows, such as those of the northwestern United States. They originate rapidly and thus stand intact as high lava plains for a long time before the valleys become much deepened.

The largest plateaus, such as the Colorado and that of Central Asia, cover tens of thousands of square miles and include entire mountain ranges and vast plains. The different sections into which they have been cut are usually named individually. (Hence "Colorado Plateau" and "Colorado Plateaus" both are encountered.) Some plateaus have the form of mountain peaks cut out of high terrain; for example, the Beartooth Plateau of Montana and the Catskill Plateau of New York. Often a plateau is a height of land between ranges, as is the Flattop region of Colorado. Some of these land masses, like the interior low plateaus of Kentucky and Tennessee, grade into the dome form.

Every continent has plateaus of great size. In eastern North America there are the Appalachian, Laurentian, and Labrador plateaus. In the West the largest are the Colorado and the Columbia, along with the lofty central

On the edge of the Columbia Plateau: The Columbia River is in the foreground, and the two large lakes beyond are Ancient Lake (*left*) and Dusty Lake.

area of Mexico. In Europe north of the Alps the Black Forest Plateau reaches into Germany and Czechoslovakia, and to the west there is the plateau of southern France. Spain consists almost entirely of a broad highland, and so do Yugoslavia, Turkey, Iraq, Iran, and Afghanistan. Northeast Africa has a broad plateau whose southward extension forms the southern third of the continent. South America has major plateaus in Venezuela, Brazil, and Argentina, and Australia has several in its interior.

Lava plateaus are generally smaller than these great highlands of uplift, but are of respectable size notwithstanding. Basaltic lavas pouring out over the crust and volcanic debris raining down upon it have covered millions of square miles of the planet. Being very fluid, basaltic lavas can travel almost incredible distances; the flow from the 1859 eruption of Hawaii's Mauna Loa reached the sea, thirty-five miles away. Flows in the Columbia Plateau region overwhelmed mountains nearly a mile high. Even rhyolitic lavas, which are acidic and quick-cooling, can spread far, as the flows on the Yellowstone Plateau and in northeastern New Mexico demonstrate. Dust from an explosive volcano such as Mount Mazama or Jemez Volcano can bury a region to depths of scores of feet.

Any extensive land surface covered to a level with lava flows, and standing above adjacent terrain, is a lava plateau. A lava plain is any similar surface that is not so high-standing and, usually, is smaller. The distinction is not always clear. It should be noted that whereas all large plateaus of uplift have been deeply dissected, extensive lava plateaus may be substantially intact.

In North America the heaviest known lava flows were those that covered 175,000 square miles or more in the Pacific Northwest during the mid-Cenozoic, 15 to 20 million years ago. In Washington, Oregon, California, Nevada, and Idaho successive basalt floods 10 to 20 feet thick almost completely engulfed landscapes with a relief of 6000 feet—including the Blue Mountains among others. The "steptoes" of Oregon are granite mountain summits protruding above the surface of the congealed flows. Most of these eruptions occurred at intervals of centuries or millennia within a span of a few million years. Some of the lava came from volcanoes, but most of it from widespread systems of fissures, and the total volume amounted to about 60,000 cubic miles. Interbedded with the basalt layers are strata of pyroclastics and consolidated dust, as well as erosion sediments laid down between eruptions. The northern half of this region is the Columbia Plateau proper, and the other areas, individually named, are collectively called the Columbia Plateaus—not a very clear distinction.

West of Spokane, Washington, occupying about 40,000 square miles of the Columbia Plateau, are the well-known "scablands." Here meltwaters blocked by the edge of a Pleistocene ice sheet at the north end of the plateau rushed southward, cutting the lava land into a weird system of canyons (coulees), mesas, and buttes. Hanging valleys on the edges of the canyons, waterless waterfalls, and thick gravel terraces along abandoned, potholed stream channels testify to a turbulent, watery past.

In eastern Washington the Snake River has cut through all the basalt layers of the plateau and into the granite basement beneath. For a stretch of 40 miles, west of the Salmon Mountains of Idaho, the canyon averages 5500 feet in depth and reaches a width of 15 miles. In size if not in colors this great gash in the land of lava ranks with the Grand Canyon of the Colorado.

Most of the Northwestern lava plateaus are well dissected, except for the Snake River Plain on the eastern side. Canyon walls often show the striking columnar structure of basalt. On rock faces thousands of feet high the profile of the buried Blue Mountains can be traced. Because of the porosity of the tuff interbedded with the basalt, there are many natural pockets for water storage, and springs are found at numerous outcrops of the strata.

Southern Mexico and Central America also show extensive lava-covered landscapes. South America has the great basalt-covered Paraná Plateau of southern Brazil, Uruguay, and Argentina. On the Deccan Plateau of India 200,000 square miles are covered with basalt flows piled up to depths of a mile or more. On North Island in New Zealand rises the Ignimbrite Plateau, some 5000 square miles in extent, formed by deep consolidated volcanic ash.

On the Appalachian Plateau: The rolling terrain around Cambridge, New York, is typical of much of the region west of the Folded Appalachians.

Well known too are the widespread plateau basalts of Iceland, some of which were produced by disastrous eruptions within historic times. Of similar origin is the 1000-square-mile Antrim Plateau of Ireland, formed by earlier outbursts of basalt from the North Atlantic's bottom.

But the greatest plateaus owe their elevations to uplift. During their rise they have been divided by stream work into smaller units which are separately named. Thus the 130,000-square-mile Colorado Plateau includes, among a multitude of others, the Kaibab, Kanab, and Shivwits plateaus. As the small plateaus in turn are dissected, mesas or mountains of circumerosion may be created.

The top surfaces of a plateau often are rather flat and may be accordant—that is, of about the same elevation. Where such surfaces can be proved to have been leveled by erosion, as seems true of the Flattop surface in northwestern Colorado, they may be the remnants of a peneplain which has been uplifted. More often the flatness and accordance of summits are due to horizontal rock structure; then the highland may be called a structural plateau. Examples in the United States are the Colorado and Appalachian plateaus. The plateaus of Germany, Spain, and Central Asia are of similar origin. Certain other heavily eroded and uplifted old lands, such as Canada's Laurentian Hills, also are called plateaus, as are some batholithic domes.

Plateaus of weak rock are likely to be areas of badland topography. The Badlands of southwestern South Dakota, for example, are the last remnants of a plateau that has been reduced nearly to the level of the surrounding plain. The rock, consisting mostly of poorly consolidated clays, has been highly vulnerable to stream erosion in the semiarid climate. The Painted Desert of northern Arizona is the remnant of another plateau of very weak rock—actually, weakly compacted volcanic dust—upon which destruction by streams has been swift.

During long erosion an isolated part of a plateau may briefly become what is called a stripped, or structural, plain. This owes its flatness to horizontal structure strictly: it has been eroded down to a resistant level stratum upon which dissection has barely begun. Such a plain is most likely to be found on an arid landscape, because in a humid region streams would cut deeply into the edges even while the resistant stratum was being exposed. Structural plains thus are comparable to hard-topped mesas: they exist thanks to differential erosion and remoteness from streams. They differ from mesas mainly in being more extensive. In turn, structural plains differ from structural plateaus in being smaller and less dissected. Numerous small tablelands of the Colorado Plateau are structural plains, and the term has been applied also to some of the platforms developed locally in the Folded Appalachians. One of the most scenic of all structural plains is Table Mountain, at the tip of South Africa—the remnant of an extensive plateau.

Plains: Undissected Flatlands

The distinction between plateaus and plains is not always clear. Usually, however, the term "plain" is favored for terrains that are quite level and have large undissected areas, and that owe their levelness not so much to rock structure as to erosion (as in the case of peneplains) or to blanketing with sediments (as with coastal plains).

A coastal plain is a part of a continental shelf that has recently emerged to become dry land either by uplift or by lowering of sea level. The shelf is the gently sloping, sediment-mantled rock surface that lies along most continental margins between the shoreline and the more steeply inclined continental slope, which leads down to the ocean bottom proper. Where coasts have sunk substantially relative to sea level, as has the coast of Maine, there are no coastal plains.

The sediments covering a coastal plain may result from marine deposition or pre-uplift shore erosion, or may come from inland via streams. The plain's near-flatness may be due partly to the leveling work done by waves and currents before the emergence. Down the gentle seaward slope, in approximately parallel courses, run consequent streams, which are met by

Becoming a coastal plain: The Everglades of Florida are on what was until recently a sea bottom. Here is a coastal plain still in early stages of emergence.

ROCK, TIME, AND LANDFORMS · 192

the subsequent tributaries at wide angles. The seaward edge of the coastal plain is likely to be fairly straight or broadly curving except where it is indented by streams.

Elevation on a coastal plain generally increases landward; and so does the relief, because of the increased elevation and because erosion inland has been going on longer. Shallow basins on the plain, produced by downwarping or by wind or wave work, may be the sites of lakes and marshes. A series of ocean-facing wave-cut terraces may be present, representing intervals of stillstand during the uplift. Where a tilted resistant stratum of the old land terminates in a cuesta with a landward dip, this will have been cut through by streams to form water gaps. Far from the shoreline, erosion may have removed the sediments and exposed older rock terraces.

North America has the Atlantic and Gulf plains and the smaller coastal platforms of northern Alaska and the northernmost Canadian islands. All have emerged within the past few million years. The Atlantic plain, covered with sediments washed and blown down from Eastern highlands during the past 100 million years, lies over the eroded, sunken Piedmont Appalachians. These are buried to depths which along the coast of Georgia exceed fifteen thousand feet. The Gulf Coastal Plain has been for

Land of the sky: The plains east of the Rockies are on bedrocks that are mostly horizontal, with little relief. This area is in the Panhandle of Texas.

tens of millions of years the dumping ground for rivers draining the central region of the continent—the Mississippi, the Rio Grande, the Chattahoochee, and others.

The 500 billion tons of dissolved and solid materials poured into the Gulf yearly by the Mississippi alone suggests how much of the continent's substance now lies on the Gulf Plain. Within a few million years this area has advanced scores of miles southward. Most of the state of Louisiana has been covered by the Mississippi's dumping. The weight of the sediments has depressed the bottom of the Gulf itself six to eight miles.

The Atlantic Plain and the Gulf Plain from Alabama to eastern Oklahoma and Texas show a belted pattern due to cuesta-forming outcrops of sedimentary strata. Along the Gulf are many salt domes, which originated from deposits on the floors of ancient salt lakes and marginal seas.

South-central England is predominantly a coastal plain—an extensive lowland which has emerged from the sea only during comparatively recent geologic time. The Chiltern and Cotswold hills represent successive cuestas facing eastward. The Irish and Scottish lowlands also are coastal plains. In France the coastal plain is known as the Paris Basin, which accounts for more than half of the nation's area. The cuestas of the Paris Basin are in the eastern zone, and they face east because of the downwarping that produced the basin.

Since the edges of the continents rise and sink intermittently, a coastal plain at any given time is likely to be either losing elevation or gaining it. The coast of New Jersey according to recent estimates has sunk about four feet during the past thousand years; but this slight sinking has followed a much greater uplift, so that the plain is still above sea level. The coastal plains of Alaska and Canada, and of Scandinavia also, are gradually rising and growing in area as the crust continues to rebound following the melting of miles-thick Pleistocene ice. Through geologic time there has been a strong net gain for emergence around the continental edges.

Alluvial Plains: Smoothed by Stream Deposits

Wherever streams periodically flood over the land, an alluvial plain is built up. Some such plains, like those bordering the lower Mississippi, are valley floors that have been given a smooth surface by thousands of years of deposition. These are essentially flood plains, showing meanders, oxbow lakes, natural levees, and other features that express low stream gradient and the dominance of deposition. They may have become so extensive that the valley sides are many miles from the center of the plain. Alluvial plains may form also where wide alluvial fans coalesce, as in the San Joaquin Valley of California, on the High Plains east of the Rockies, in France north of the Pyrenees, and in parts of the Po Valley in Italy. Wide coastal areas also may be essentially alluvial plains.

Still another variety of alluvial flatland is the delta plain, formed by flooding distributaries. A delta plain differs from other alluvial plains essentially in the fact that it is built up on a sea or lake bottom. It may be so nearly flat as to look like the surface of a calm sea, but it may have hollows occupied by shallow lagoons. Delta plains tend to be arcuate or fanlike in outline, as is the Nile Delta; the bird's-foot outline, like that of the Mississippi Delta, is uncommon. A delta plain formed by an estuary takes the shape given by the confining valley walls.

Bajadas, Bolsons, and Lake Plains

Among arid highlands where runoff is torrential but only temporary, an alluvial fan is commonly seen at the foot of each ravine or valley. Such fans often grow to such size that they coalesce to form alluvial plains.

In desert country where drainage is especially poor, as in the Basin and Range Province, the alluvial plain may extend two or three miles from the base of the highlands and then grade into the sediment-covered, very gently sloping surface known as a bajada, which in turn leads down to a central basin. This basin, produced by faulting, is called a bolson (Spanish, "purse"). Having no outlet, it is base level for the area circumscribed by the highlands; it is the regional sump. Here rapid runoff periodically forms a temporary shallow lake—a lake that may last hardly longer than the shower that creates it. As filling of the bolson by mineral-bearing runoff and mass wasting continues, its surface approaches the flatness of a true plain, and it becomes a playa.

The playa is but one variety of lake plain. There are as many other varieties as there are lakes: glacial, delta, coastal-plain, blocked-valley, and so on. Most are formed by accumulations of sediments in abandoned basins, the sediments being made up of rock waste with remains of plants and animals mixed in. Once a lake has dried up, wastes soon fill the basin, and before long only a trained eye can see that it was once the site of a lake.

Far-sprawling plains are seen today where meltwater formed lakes along the edges of the Pleistocene ice sheets. Lake Agassiz, the greatest of all such bodies of water, left a plain 700 miles long and as much as 250 miles wide, reaching from northern Minnesota and North Dakota northwestward into Canada. The flats of Lake Bonneville, Pleistocene ancestor of Great Salt Lake, cover nearly one third of the state of Utah. The Basin and Range Province is much pocked with abandoned lake basins. Thousands of small lacustrine plains are seen in the region from northern Minnesota to Maine and northward to Hudson Bay. Small lacustrine plains of Pleistocene vintage formed also on valley bottoms among glaciated highlands, across northern North America and northern Eurasia.

A bolson plain: The floor of Death Valley, here seen with the Panamint Mountains in the background, has been leveled by stream deposits. The view is toward Badwater.

Glaciated Plains: Smoothed by Ice

Landscapes broadly smoothed by glacial abrasion are called ice-scoured plains, and those leveled by the spreading of glacial debris are till plains. All the flatlands covered by the Pleistocene ice sheets fall into these categories. They represent millions of square miles of North America and other continents.

Ice sheets grinding southward over central and eastern Canada scraped away most of the soil cover, and planed and gouged the bedrock to depths approaching forty feet. Canada's ice-scoured plains show thousands upon thousands of square miles of bared bedrock, dotted with uncounted rock-basin lakes. Even more extensive flatlands are covered with glacial debris, for the Pleistocene ice did perhaps the greatest bulldozing operation of all time. Thousands of cubic miles of soil and broken rock was carried or pushed down into the United States, and flooding meltwaters spread it far

to the south. This material, called glacial till, was dropped on the north-central plains, especially in Illinois and Indiana, thickly enough to bury hills several hundred feet high. Till plains in these states, and to some extent in neighboring states, spread over thousands upon thousands of square miles.

A till plain in the Midwest: Farmland near Morley, Iowa, consists largely of glacial till spread widely by meltwater from a Pleistocene ice sheet.

IX
Realms of Wind

Barchans in Saudi Arabia: Plenty of sand is available in the Rub' al-Khali area for the building of large dunes. The climate is so arid that little vegetation can develop to hinder the work of the wind.

To the stark dryness of the desert world is owed a special gallery of landforms. In this place of far horizons are sand seas and ranks of marching dunes, empty river beds and sprawling alluvial fans, glistening playas and polished rock pavements. The grotesque pinnacles and twisting chasms of badland topography are here, and ragged mountains half buried in rock wastes. Such features, though not present in all deserts and not restricted entirely to desert terrain, well express the desert's nature.

Deserts are defined as lands with less than ten inches of rain per year. They make up nearly a third of the world's landscapes, including some areas which, as in Central Asia, are notable more for cold than for heat. Among all terrains, deserts are surely the most paradoxical, because although they are dry their landforms are mostly the work of torrential streams. Here, perhaps more than on any other terrain, running water is the master sculptor.

It is the global pattern of land and sea that accounts mainly for the desert's aridity. The sea, as we recall, is the source of most of the water that falls as rain and snow. As moist sea air moves over land, it is either warmed or cooled. If it is cooled, its ability to hold water as vapor diminishes, and precipitation is likely. If it is warmed, the chances of precipitation are reduced. Such are the basic conditions that make some lands humid and others dry.

The west-coast deserts of middle latitudes, such as those of the Basin and Range Province, owe their aridity to nearby mountains. As the moist air of the sea winds climbs the coastal ranges it expands, because atmospheric pressure decreases with elevation. Expansion cools the air enough to condense much of its moisture as rain or snow, which falls mostly on the mountains' windward slopes. When the air slides down the leeward slopes, it is warmed again by the increase of atmospheric pressure and turns extremely dry. Water for the leeward regions, those in the "rain shadow," must come from an occasional shower produced by unusual conditions, or from rain- or snow-fed streams coursing down from the highlands.

This is the pattern in the western deserts of North America from Canada to Mexico and in the Atacama Desert of South America. The latter region, located east of the windbreak formed by the Andes, is the world's driest terrain. According to actual measurements, it has received as little as five one-hundredths of an inch of rain over a twenty-five-year period. The desert of Morocco, guarded by the Atlas Mountains of Africa, is almost as dry.

Three great deserts in Central Asia—the Gobi, the Tarim, and the Turkestan—owe their dryness to invasions by cold, dry air from the polar regions and to their remoteness from oceans. Air moving down into Central Asia from the north grows warmer and thus drier. Air moving up from the Indian Ocean, to the south, loses its moisture in the ascent of the lofty Himalayas. Winds from the west suffer similar drying in their prolonged journey from the Mediterranean and the Red Sea.

A different pattern prevails over the subtropical deserts: the Sahara and Kalahari of Africa, the Arabian Desert, and the vast central desert of Australia. All these lie in the horse latitudes, which are zones of high atmospheric pressure where the typical weather is governed by calm, sinking air rather than air climbing over mountain masses. The trade winds that sweep these lands are moving equatorward. The warmer they get, the drier they get; and the farther they travel from the sea, the more arid they make the land.

A special category of desert is represented by the ice caps of Antarctica and Greenland and by the bleak, rocky terrain that is exposed through them and around them. Since the water here is mostly frozen, evaporation and precipitation are minimal. Vegetation may be sparser here than in the driest of the warmer deserts. Streams of water are of little importance as land-carvers, and the wind, finding little material of small caliber to move, is likely to be less effective in erosion here than in warm deserts.

The look of most deserts is explained not only by aridity but also by poor drainage. In the Basin and Range Province, for example, diastrophism has produced basins in which rainwaters and meltwaters become trapped. Precipitation is inadequate to overflow these basins; hence streams do not flow outward from the region. Instead, the water soon vanishes by evaporation and absorption into the sands. In some other deserts drainage is poor simply because the elevation is low and the streams lack gradient. By contrast, on the high arid terrains of Greece and Yugoslavia, where rain is

A desert streambed in southern Arizona: When snows are melting fast or heavy rains are falling on the Tucson Mountains, the ravine fills with a roaring torrent. The rest of the year it is dry.

plentiful, surface streams are weak because water from rain runs quickly underground through openings made in the limestone bedrock by solution. All these dry lands may look unkempt because, drainage being poor, the rock wastes accumulate in large quantities.

The Desert Stream

In humid localities the rivers normally receive enough runoff and ground water to maintain their flow seaward and to grow in volume as the sea is neared. But desert runoff is infrequent, the water table is usually far beneath the streambeds, and evaporation from streams is swift. Hence most desert rivers are intermittent; their beds contain water only for brief periods during or after rains or during the melting of highland snows.

The few desert streams that are permanent have their origins in lands of generous rainfall, and still they must be mighty to get through the desert alive. The Nile, rising in the Abyssinian highlands of Africa, enters the Sahara Desert in tremendous volume. Seepage and evaporation are severe, but enough ground water is added so that the river can complete the 1200-mile journey to the Mediterranean. The Rio Grande and the Colorado, in the United States, likewise manage to survive desert courses. Some sizable desert rivers empty into salt lakes or sinks—for example, the Owens River, which empties into Owens Lake in California; the Humboldt of Nevada,

Alluvial fans below Big Devil Rock, Mohave Desert: Deposits are thick here because streams are only temporary. The fans merge into a bajada.

Salt deposits: Devils Golf Course, in Death Valley, was the bottom of Lake Manly during the humid Pleistocene. The lake became salt as it shrank, then vanished entirely.

leading into Humboldt Sink; and the Shari River of Africa, which ends its career in Lake Chad.

Desert rains, when they do come, fall mostly upon highlands in the desert or near it. The highlands lack sufficient vegetation to absorb or slow the runoff; also, for reasons soon to appear, they are likely to be steep-sloped, so that runoff is swift. The water plunges down the slopes in sheet-floods or as torrents in gullies. It may be so swift as to cut gullies straight down, regardless of variations in rock resistance. Desert streams are so often torrential that the amount of rainfall in desert storms tends to be exaggerated.

Toward the foot of a typical desert mountain ravine, a stream abruptly loses speed as the slope eases. Dropping large rock fragments first and then the material of smaller caliber, the water spreads out to form an alluvial fan or cone. Deep channels called washes may be cut into a fan by vigorous branches of the stream. Much water is absorbed by the fan, and some of it may emerge farther down as springs. As each rain or period of melting in the mountains adds rock waste to the fan, its lower edge advances farther into the valley or basin.

Beginning just above the upper edge of the alluvial fan and reaching down beneath it there is sometimes a rock fan. This is an area of bare or thinly covered bedrock at the point where the ravine's slope suddenly eases. The rock fan apparently results from severe scouring as debris-laden torrents strike this area before spreading out over and into the material below.

The rock fan may merge into a rock pediment, which is a gently sloping bedrock platform covered thinly by the upper part of the alluvial fan and much more deeply by sediments at lower levels. Exactly how rock pediments originate is a matter of controversy, but erosion by flooding streams—in an earlier, perhaps more humid time—is a likely cause.

Along the base of a highland one may see a row of alluvial fans, their lower portions merging to form a broad, very gently sloping accumulation of erosion debris called a bajada. The bajada, usually hundreds of feet thick, extends like an apron from the base of the fans, reaching with gradually diminishing gradient out to the bolson, or central basin for the local drainage. Rainwater, running down over the bajada in sheetfloods or through shallow, frequently changing channels, eventually reaches the bolson to form the familiar playa lake. Such a lake is highly charged with minerals because it lacks an outlet; each new influx of water brings in more mineral matter, but only the water evaporates. These lakes are called salinas if common salt predominates among the minerals.

If a desert terrain is high-standing and the exposed bedrock is very weak, stream erosion may produce badlands. Here gullying rather than sheet erosion is dominant. Gullies are closely spaced and deep because downcutting in weak material by streams of high gradient is so very rapid relative to weathering. Badlands such as those of the Dakotas, Wyoming, Montana, and the Painted Desert in Arizona are spectacular examples.

Badlands are laced with systems of deep, zigzagging canyons. These are often called box canyons because a person on the bottom seems to be completely enclosed by the precipitous walls. Such canyons usually have been cut into belts of weaker rocks, and during rain the canyon rims, formed by edges of the more resistant rock strata, may become waterfalls. Often these canyons fill with sudden, muddy torrents of great depth and violence. Most of the water is soon lost in the dry canyon mazes, but there may be enough to race down through the washes and flood the valley or basin. Many a desert traveler has been caught in a wash by a wave of water from a rainstorm in the mountains.

Saturation of the typically abundant rock waste on a desert slope often starts a mudflow. Such flows are common in badlands and on arid highlands where the rock waste gets slippery when wet, as do clay and the lava dust that lies thick on some old volcanic areas. A mudflow is intermediate between a landslide and the flow of a river. Moving at a speed of several miles per hour, it can take a desert town by surprise and completely flood the streets. Mudflows act as canyon-fillers in badland regions and also contribute large amounts of sediments to bolsons.

It was long the custom to say that wind, not water, is supreme in the desert. Wind is indeed a power in arid lands, far more than under a humid climate, but running water dominates. In fact, much of the dramatic "erosion scenery" of deserts of our time—the ravines and canyons, the alluvial fans and rugged highland summits—dates from the rainy interludes of the Pleistocene. Judging from the mountain valleys and the multitudes of now dry lakes in the Basin and Range Province, the streams of yesteryear in this locality were full and powerful.

Dry-Country Weathering

In the desert, things that are quite perishable elsewhere—old guns and knives, pieces of clothing, and bones and other remains of animals—are long-lasting. And so are rocks. Because of the extreme dryness, chemical

"Elephant backs": At the western edge of the South Dakota Badlands the typical erosion forms in the weak clays are broad and rounded.

decomposition and solution of rocks are slow. With thin vegetation there is little rock destruction wrought by plant roots and by chemicals deposited by plants on the rocks. Water being scarce, frost prying becomes a minor factor, unless the desert happens to be in high latitudes. Some destruction of rock surfaces by quick temperature changes does occur, but not to the extent that tradition says.

The nature of weathering here helps to account for the generally angular appearance of desert landforms—the steepness of cliffs, the depth of canyons, the ruggedness of hillsides. Because weathering and mass wasting are so slow, streamwork though intermittent is relatively vigorous, and valley walls are kept steep. Their steepness may be accentuated by the tendency of floods in the streambeds to spread out and undermine the valley walls. Some steepening may be due also to the work of the wind, which, as erosion opens up the valleys and isolates the highlands, blows even harder through the region, sandblasting the lower portions of cliffs most severely because it cannot lift sand more than a few feet into the air.

Wind Work

The wind works in desert places with a curious skill. Because of the poor drainage, there may be plenty of erosion debris for the wind to use in sculpturing rocks and building dunes. The sand, unprotected by vegetation, is easily moved unless covered by or mixed with consolidated material, such as crystallized salts accumulated on a playa. The wind keeps blowing rock particles against the exposed bedrocks, polishing them and, in so doing, making more rock waste. At the same time the waste already on the ground is constantly reworked, the particles becoming rounded and reduced to smaller and smaller sizes. When small enough, they are blown out of the desert as dust, but the sand is left.

Wind-borne materials usually have a high content of quartz—that very strong mineral which is one of the commonest constituents of surface rocks. Small bits of quartz can have excellent cutting edges. An air current rising at a speed of 2 miles per hour can lift a bit of quartz $1/250$th of an inch in diameter, and a current rising at 27 miles per hour can carry with it grain $1/20$th of an inch in diameter. As in a stream, bits of gravel too heavy to lift can be bounced, rolled, or slid along the ground, and larger objects can be undermined. An average dust storm is believed to contain up to 4000 tons of dust per cubic mile. Such figures show that the wind has a very respectable carrying power and that its abrasive work must be tremendous.

The effectiveness of this work near the ground is suggested by the shaped rock fragments variously called faceted pebbles, ventifacts, and dreikanter (German, "three-sided objects"), found in windy, very sandy areas as different as the eastern Sahara and Cape Cod. These fragments, some resembling Brazil nuts, have relatively flat, triangular facets produced solely by abrasion. Stones long exposed to a prevailing wind without change of position may be beveled to a point on the windward end, while those that have been disturbed from time to time or subjected to variable winds develop many facets and perhaps more than one point.

Wind work slackens sharply above an elevation of about four feet; hence the form of the desert's mushroom rocks and pedestal rocks, which are broadest at the top. Often these rock figures have been carved out of horizontal strata of varying resistance, which are cut away at different rates and thus produce irregular profiles. Wind eddies may cut caves into the base of cliffs, and windows and arches may result from the wearing through of narrow ridges. In much of this sculpturing, rainwash may dominate but the wind's work is undoubtedly important.

Sand grains beating against a rock face, century after century, find all the points of weakness and etch the surface in intricate detail. Resistant minerals in the rock, such as quartz veins or clusters of quartz crystals, are left as raised patterns. In cliffs of sedimentary rocks, differential destruction of the edges of the strata makes a fluted pattern. Where outcrops are

Desert windows: Many openings such as these at Arches National Monument, Utah, have been produced by underground stream cut-offs. Some may result from the weathering-through of thin, finlike divides that separated former canyons.

tilted, undercutting sometimes leaves overhangs with wavy, cockscomb-like edges—the yardangs typical of Asiatic deserts. In all these sculptures, too, rainwash is allied with the wind.

Desert pavement (called *serir* in the Sahara) is a natural mosaic platform of rock fragments that have been too heavy for the wind to move and have been smoothed and leveled by blowing sand. A high polish, called desert varnish, may gleam on these rocks, especially those coated with oxides of iron or manganese that have risen to the surface in solutions by capillary action. These polished pavements are kept clear because the sand sweeps so easily over them.

Desert pavement is seen often where erosion debris of large caliber has accumulated, as near the foot of an alluvial fan. But it may be encountered on any desert area from which the small-caliber waste has been mostly blown away. Such areas develop where the degradation process is advanced and the terrain has become fully opened to the sweep of winds.

In some localities the sand-charged wind has scoured its way down through the pavement to bedrock. Thus is produced the common rock-floored plain, which in Egypt is called *hammada*. Often it is covered with pebbles. Although "desert" means to most people a sandy terrain (the *erg*

Desert pavement: These rock fragments in Death Valley, too large to be moved by wind, remain while sand is being blown away. Mineral-bearing waters consolidate them.

of the Sahara), desert pavement and hammada are actually more common on deserts than sandy areas are.

When the wind blows very hard across a dry terrain covered with light soil or sand, the notorious dust storm or sand storm results. In a dust storm the load may be lifted thousands of feet and transported over the breadth of a continent. In a sand storm most of the load travels low, moving by saltation (picking up, carrying, and dropping of the grains) and by surface creep, which consists of slight forward movements of particles due to impacts of other particles blown against them. For reasons already mentioned, a sand storm is highly erosive near the ground.

The formations most vulnerable to wind erosion are clays and shales. The rock grains are often poorly consolidated and easily loosened; and, being fine and light, they are easily lifted high and carried far. Clouds of this dust may become so dense as to threaten man and beast with suffocation—not to mention the burial of crops. During the 1930s winds from the Dust Bowl of the south-central United States darkened the skies over the East Coast, and in some areas deposited as much as thirty-five tons of dust per square mile. But such events were not unprecedented. Since ancient times a wind of the Sahara Desert, known as the sirocco, has been blowing

A blowout in Harney County, Oregon: Vegetation has anchored some of the sand in an area where winds have reduced the landscape by 4 to 6 feet.

red dust as far north as Germany, causing "red rains" en route. Africa is slowly being added to Europe.

Here and there, wind-scouring makes the basins called deflation hollows, or blowouts. These develop where the wind is strong and the rock waste is loose or the exposed bedrocks are weak. As the wind cuts deeper and deeper, the water table may be reached, and the moisture may anchor the rock waste against further blowing. In such hollows are found many of the oases of Egypt and Libya and the "pans" of South Africa and the Kalahari Desert. Some oases are spots where water under hydrostatic pressure comes to the surface through a fault or between tilted strata.

Deflation is often a problem for man. Winds blowing over the Nile Delta during the past 2600 years have removed an estimated eight feet of the fertile soil. Even in more humid climates, such as that of the south-central United States, deflation becomes a grave farm problem, especially during droughts.

Wind-scouring is certainly responsible in part for the low elevation of many desert areas and the resulting poor drainage. One such area, perhaps, is the Qattara Depression, some three hundred miles east of the Nile Delta. It is 180 miles long, averaging 50 miles in width along most of its length, and reaching 420 feet below sea level. This has been interpreted by some geologists as a blowout and by others as a graben; perhaps it is both.

The blowout called Big Hollow, near Laramie, Wyoming, is 9 miles wide and 3 miles long, and 300 feet deep on the average. From it something like ten billion tons of material has been blown out. This hollow is the largest of many, some containing lakes, that occur in a broad belt across the High Plains of the United States from Texas to Montana. Another large Western blowout is Danby Playa, in the southwestern part of California's Mohave Desert; its host of very small, scattered mesas are unusual.

It is mainly by wind work that the world's supplies of loess have been scattered so widely. Loess is a yellowish, fine-grained material produced mainly by the abrasion of bedrock by glaciers and desert winds. Most loess deposits are now found to the leeward of the great deserts and of areas glaciated during the Pleistocene. Thick accumulations on the central United States have come from the Rockies and the Western deserts.

Dunes and Other Wind-Blown Accumulations

Clay particles are blown away from deserts, but most of the sand remains. Where sand covers the bedrock but thinly, or is mixed with cobbles and gravel, it is likely to be distributed as thin rippled sheets. Where wind blows it against a cliff or a wall, sand drifts are built up at the foot of the obstacle. These drifts are seen also as pile-ups behind a gap in a barrier. The term sand shadow is used for a longitudinal ridge of sand drawn out by

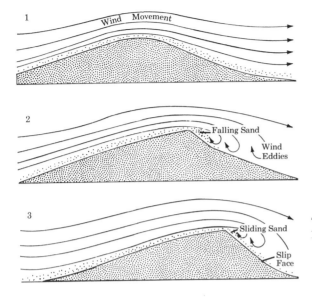

The making of a barchan: Dune forms express the resistance of abundant sands to removal by the wind.

the wind behind a narrow obstacle like a fencepost, or a deposit of sand that has been blown over a cliff or other steep slope.

Of all the shapes taken by sand deposits, dunes are the most picturesque and elaborate. They form most readily on flat terrain—not only in deserts and along sandy seacoasts, but wherever there is plenty of sand not anchored by vegetation or heavy-caliber sediments.

The barchan, typically crescent-shaped, starts building where blowing sand encounters a minor obstacle, or where the wind is not strong enough to transport all the loose sand that has accumulated. As a deposit builds up, it becomes itself an obstacle and keeps growing by acquiring the sand it intercepts. Grains moving up the front of the sand heap by saltation are dropped on the back side, or slip face, and slide down, forming a slope of about 30 to 35 degrees—the angle of repose for dry sand. The slip face becomes hollowed out by the action of eddies. Meanwhile, as the wind meets resistance in front, it tends to be deflected toward the sides; and because less resistance is met there, the sand particles are moved further so as to extend the sides. Thus the barchan acquires its leeward-pointing "horns."

A well-developed barchan is highest at the center and convex toward the wind, and the ratio of width to height averages about twelve to one. The size is limited by wind conditions and sand supply. Where the wind is strong and sand plentiful, as in the erg of the Sahara, barchans achieve heights exceeding five hundred feet. They tend to occur in swarms, and their orientation shows the direction of the dominant winds. Barchans

Barchan dunes: These are characteristic at White Sands National Monument. Wind carries the sand skittering around the edges of one dune, then up the slopes of the next. Sand that flows over the center of a dune is trapped on the steep slip face. When it has built up to an unstable angle at the brink of the slip face, it avalanches down the slope, and thus the dune moves forward.

Short sections of transverse dunes: The ridges at White Sands National Monument are seldom straight for long distances, but sweep back and forth as troughs and crests speed and retard the flow of the sand across the dunes.

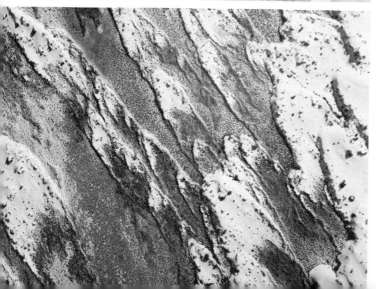

Parabolic dunes: Abundant vegetation in this part of White Sands National Monument acts to hold the shallow parts of the dunes in place. Only the sand in the center of the dune is free enough to be moved onward by the winds. The result is the development of a characteristic parabolic or U-shape.

crowded together may coalesce into the more irregular structures called transverse dunes.

Where vegetation such as bunch grass gets established on a dune, it slows the wind. The dune may thus become anchored, though it can continue to grow. Unanchored dunes are likely to migrate forward as long as the wind continues blowing predominantly in that direction. Migration occurs as sand continues to be moved up the windward slope and dropped on the slip face. Movement averages twenty feet per year for large dunes and more for small ones, but rates as high as two thousand feet annually have been observed where winds are strong and sustained.

In desert country, some bolsons and canyons have become filled by migrating dunes, and whole towns have been covered by them. On coastal plains the protection of houses and other structures against marching sand is often a grave necessity. With time, dune migration results in the deposit of enormous amounts of rock waste on humid lands beyond the desert's borders.

Contrasting with the barchan in form and usually not found on the same terrain is the seif, well known in the interiors of the Saharan and Arabian deserts. This is a long, sharp-crested, and sharp-pointed ridge (hence *seif*, Arabic for "sword") built up parallel to the prevailing winds, the ratio of width to height being about six to one. In the Sahara seifs reach heights of three hundred feet, and in Iran, six hundred feet; lengths of

Seif dunes in Saudi Arabia: These dunes, in the Rub' al-Khali area, are formed by processes not well understood. Sometimes they attain lengths of hundreds of miles.

Dunes on Cape Cod anchored by vegetation: The migration of these sand hills near Provincetown, Massachusetts, has been slowed by plantings of beach grass.

hundreds of miles are not uncommon. Seifs generally occur in groups, often with alleys of bare rock, called gassi, showing between them. Once attributed to very strong winds blowing constantly in one direction, seifs are now thought by some observers to be barchans strongly modified by cross winds. The sharp crests on seifs often suggest barchan forms.

The largest of all dunes are the whalebacks of the Sahara. These round-backed monsters surpass even the seifs in bulk and length. Like seifs they develop longitudinally, parallel to the dominant winds.

Where winds are variable or the terrain is uneven, dunes may not develop at all, or they may fail to maintain regular shapes. Various combinations are encountered, especially the incipient barchan form. But variations can be due also to interference by vegetation. On the Navajo Desert of Arizona, for example, and at White Sands in New Mexico, one finds the parabolic dune, a sort of reversed barchan, with the windward side hollowed out and the horns pointing to windward. These dunes may be

REALMS OF WIND · 215

barchans modified by wind under vegetation control. Comparable dunes are common along coastlines and on river flood plains in semiarid localities.

The region of greatest dune development is Saudi Arabia, about 400,000 square miles of which is buried in sand. Next is the Sahara region, with about a third of its 900,000 square miles covered with sand. Central Asia has enormous sandy deserts, and so does Australia. In Europe the principal dune areas are along the Bay of Biscay in France, the North Sea coasts of Belgium and the Netherlands, and the Baltic Coast of East Germany. The west coast of Scotland also has areas of well-developed dunes.

In the United States the Atlantic Coastal Plain exhibits dunes spottily from Cape Cod to Florida. Most impressive are the great sand hills of Cape Cod and those of Cape Henry, Virginia. At Cape Henry giant dunes have marched inland to engulf a forest.

Vegetation-covered dunes of light sands dropped by the melting Pleistocene ice sheets are seen in New York north and south of the Adiron-

A giant dune front at Great Sand Dunes National Monument: The San Luis Valley, a graben in south-central Colorado, opens toward the southwest, from which prevailing winds blow. Through the millennia huge amounts of sand have been trapped here.

Dunes advancing through a woodland: Dunes in this part of Great Sand Dunes National Monument have overwhelmed the trees, killing them, and then moved on.

dacks. Hills of glacial sands scores of feet high have built up by winds in the vicinity of Indiana Dunes State Park at the foot of Lake Michigan. Sand hills and dunes of lighter material are scattered over alluvial plains and fans of the Great Plains from Nebraska to Texas, and appear also along the Columbia and Snake rivers in Washington and Oregon.

Dune development in North America has been greatest in the Basin and Range Province and the Colorado and Mohave deserts of California. Of special interest are Death Valley, California, and Great Sand Dunes National Monument in the San Luis Valley of Colorado. Here large grabens act as traps for sands blown from the south.

White Sands National Monument, in southern New Mexico, is unique with its dunes of glittering white gypsum. The mineral particles are carried to the dune area by dominant winds that blow across the floor of a nearby dry lake, where gypsum was deposited during the Pleistocene by streams coursing down from the mountains.

Desert Evolution

Some geologists view the various types of deserts as representing various stages of a desert cycle. The cycle is at best a highly theoretical concept, but it does make an interesting frame of reference for the deserts of our age.

The cycle may begin with the rise of highlands which drastically reduce precipitation on lowlands to leeward. Erosion on the highlands is rapid, and the debris is swept down onto the lowlands. There most of it becomes stalled, because the streams are weak and the wind's carrying power is restricted by the highlands. As these are reduced, the lowland surfaces are aggraded, and the scene gradually becomes one of a land half-submerged in rock waste.

As downwasting of the highlands continues, winds sweep in over the accumulated erosion wastes with growing power. This debris is abraded to smaller and smaller sizes, and is swept at an increasing rate toward the

Desert goblins: Mushroom rocks, typical of desert scenery, make a spectacle in the Valley of the Goblins in Wayne Wonderland, southeast Utah.

Old desert mountains in western Saudi Arabia: These are half-buried in debris. Faults cut across the strata at low center. Dikes are visible near the center.

leeward edge of the desert and beyond. Meanwhile the highlands, being lower, are less effective in taking the moisture out of the winds, and more moisture therefore falls upon the lowlands to leeward. Stream erosion gradually links the basins and, with the aid of winds, cuts outlets between the shrinking highlands, so that rock waste can be moved to lower base levels beyond the desert. The integration of drainage may mean more rapid downcutting, so that the topography at this stage becomes more rugged. Also, with erosion debris being exported faster than it is being produced, the desert may now show more desert pavement and bare bedrock floors.

As the highlands shrink further, the apronlike pediments around them grow wider and wider, in some places joining between the highland masses. Merging alluvial fans lead down to widening bajadas. The landscape now has low relief like that of the Mohave Desert of California or the Great Sandy Desert of Oregon.

In this old-age phase the once lofty highlands have become a mostly bare, gently rolling, wind-scoured landscape with perhaps a few scattered inselbergs.

Like any other landscape, a desert may become rejuvenated. Either a new uplift or an increasingly humid climate, or both, can be the rejuvenators. The new cycle may be evidenced by stronger gullying and ravine-cutting on pediments and bajadas. Fresh fault scarps may appear, and the bases of highland blocks may show slickensides. At the foot of ravines new alluvial cones and fans form, and the surfaces of old ones show signs of fresh or more rapid deposition. So the desert terrain may become young again—as, perhaps, it has been several times in the past.

X

Ice on the Land

The crystal scenery of glaciers makes a great impression on travelers, but few are aware of how much ice there is on the land, how efficiently it sculptures rocks, and how widely this sculpturing has been done in the past. Valley glaciers and ice caps today cover a tenth of the continental surfaces, mostly near the poles. During the Pleistocene they mantled nearly a third of all land, including the northern parts of North America, Europe, and Asia, and high mountains of South America and Africa as well. Most of the glaciers are now gone and the lands they overwhelmed are open to the sun, but the few millennia since the great meltings have not obliterated their handiwork. Temperate lands, as well as the far north and the far south, display horn peaks and smoothed hills, reamed-out valleys and deranged rivers, rock-basin lakes and debris-littered plains. In geologic perspective, the Ice Age ended but yesterday.

Only for about a century has the full magnitude of the Pleistocene glaciation been known. Mountain people long ago, noticing periodic ad-

Glaciers in Alaska's Coast Mountains: The view is northeast across the Canadian-Alaskan boundary. From the ice cap, valley glaciers radiate in all directions.

vances and retreats of valley glaciers, must have read the meaning of striated rock surfaces, boulder fields, and heaps of sand and gravel on ice-free lowlands. But an adequate interpretation of such signs came only with the age of geology. In 1829 a Swiss, Ignaz Venetz-Sitten, startled the learned minds of the day by announcing that all northern Europe had been glaciated. He had recognized the evidence on lowlands near the mountains of Scandinavia and central Europe, and on Europe's northern plains. Other geologists, at first skeptical, were convinced when they saw that the plains were indeed strewn with fragments of rock different from the local rock—"erratic" fragments carried down from the north by glaciers. Meanwhile the Swiss naturalist Louis Agassiz, having compared glacial features in the Alps with the evidence in northern Europe, began to talk of a recent great Ice Age. Visiting Great Britain and North America, he found similar glacial relics there. Some scientists argued that the scattered glacial debris, which they called drift, had been dropped by southward-drifting icebergs during a recent submergence of the continents, but the evidence for submergence was weak. By midcentury the Ice Age was accepted as fact.

It is known today that the Pleistocene Epoch spanned at least 600,000 years, perhaps more than twice that, and included four glacial episodes. During the last episode, which began about 70,000 years ago, ice sheets in North America reached as far south as northern New Jersey, southern Ohio, and Montana. In Eurasia the ice rode down into southern England, northern Germany, and northern parts of the Soviet Union. Local glaciers formed on the mountains of New Zealand, Australia, and even equatorial South America and Africa. Since this last glacial maximum the world's climate has become warmer by about 15 degrees Fahrenheit, and all the great ice caps, except those on Greenland and Antarctica, have disappeared. Outside polar latitudes only a few small ice caps survive, and these with most mountain glaciers are melting back today.

Of the world's estimated 326 million cubic miles of water, about 7 million cubic miles is now locked in glacial ice. Most of this ice is in the Antarctic and Greenland ice caps; only about 50,000 cubic miles is in glaciers elsewhere. All the world's ice represents enough water to cover all lands to a depth of between 100 and 150 feet.

The Valley Glacier

Glaciers form only on land of low gradient, only where snowfall is abundant, and only above the snow line—that is, where annual snowfall exceeds annual evaporation and melting. Such conditions exist on cold highlands not far from the sea. As moist winds sweep over these highlands they are chilled, and much of their moisture condenses out as snow. Accumulations of this snow make glaciers.

Snow collects in cold highland hollows, especially those sheltered from strong winds and from the sun. Accumulation occurs by direct fall, by avalanching, and by wind. During warm periods some snow melts, but it refreezes as granular masses (firn) when the temperature goes down again. This process of "nivation," with the process of compaction as new snow layers are added on top, converts the snow mass into ice. At a depth

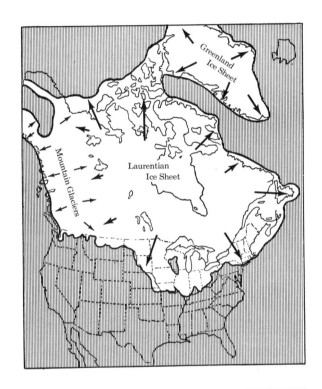

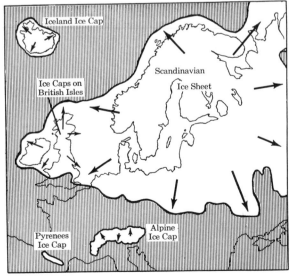

Pleistocene glaciation in the northern hemisphere: Ice movements were not all toward the south, but rather in various directions from centers of accumulation.

of 90 to 170 feet (the higher figures apply in colder zones where melting is limited), the conversion is complete.

Under the weight of overlying ice and snow, the ice near the bottom of the mountain hollow is semiplastic. As new snows fall, the ice bulges over the edge of the hollow and creeps downslope. If the slope is broad and smooth, the ice may descend as a sheet, but usually the ice drains into a valley below the hollow, and down it goes as what is called a valley glacier. Thus occurs a continuous slow-motion process in which the rate of snow accumulation on top, the rate of change of snow to ice, and the rate of ice departure from the mountain hollow are in balance. The flow of blue-white rock (and ice actually is rock) may reach a mile downslope, or ten miles, or even a hundred, before ablation—that is, evaporation and melting—destroys it.

Just how ice flows is debatable even after a century of study. Some motion occurs as slippage between layers of molecules that make up the ice crystals. Crystals, becoming lined up by the pressure like a pack of playing cards, slide over one another. Movement results also as ice particles melt under pressure and the water, meanwhile acting as a lubricant, flows to zones of lower pressure and recrystallizes there. Perhaps some motion occurs as air is squeezed out of the ice. Finally, strains in the ice break it into slabs and blocks, which gravitate downward as separate units.

In the glacier the motion is slowest at the sides and bottom, because of friction there, and fastest at midstream, where friction is minimal. The flow tends to be faster deep in the glacier, where the ice is somewhat plastic, than near the surface, where it is more rigid.

The rate of over-all motion depends upon conditions in the valley. A steep gradient favors speed. Where the channel narrows, the ice must move faster. As the flow crosses a depression in the valley floor, the ice at the bottom must stretch or speed up, while ice at the top must contract or move slower. A foot per day is a common rate for valley glaciers, but some move only a fraction of an inch daily, and others, yards.

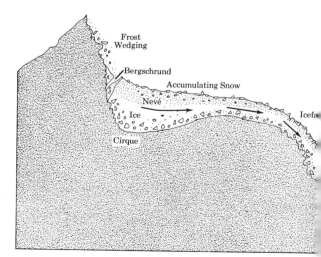

Profile of a valley glacier: The existence of the glacier depends upon an excess of accumulation over melting and evaporation at its origin.

ICE ON THE LAND · 225

Tensions in the ice produce networks of deep cracks called crevasses, extending to a depth of 150 feet or more. Tension may develop from friction between the ice and the valley walls, the passage of the ice over ledges and knobs, or widening of the valley and the consequent tendency of the ice to spread. Multitudes of intersecting crevasses may form a forest of shifting ice pinnacles, or seracs—features much feared by mountain climbers. Where the ice stream moves across a hollow, compression occurs in the upper portion, and the ice there may be fractured and heaved up to form pressure ridges.

In glacier country frost-wedging is vigorous, and fragments of weathered rock keep falling down onto the back of the ice stream. This litter, along with debris from the glacier's bulldozing activities, is carried valleyward. As the ice reaches the zone of ablation, patches of rock waste may cause uneven evaporation and melting, so that ridges and pinnacles are left. Occasionally one sees an ice pillar supporting a hatlike rock which has protected it from the sun.

At intervals in its downward course a large valley glacier is joined by tributaries. The smaller ice stream comes hard against the mainstream, causing much noisy breakage, and then flows along beside it without mixing in. The rock waste of the tributary is added as a new, separate band to the other bands of rock waste on the main glacier. By counting the bands on the mainstream one can tell approximately how many tributaries have fed it.

Far down on the mountain slope or on the lowland, a valley glacier enters the zone of ablation, where evaporation and melting reduce the ice faster than it arrives from above. From the upper boundary of this zone the volume of the glacier diminishes, with increasing signs of decay. The ice, mantled with rock debris, is in a melting state during much of the year, and its dirty, wet, pocked surface hardly suggests the clean, windswept realm of its origin.

By the time glacial ice has reached the lowland, it may be so littered

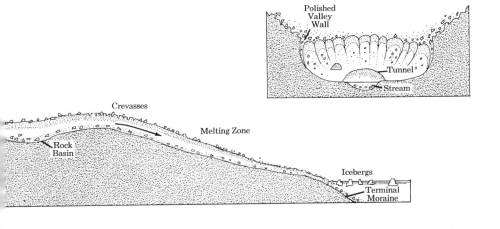

Glacier's end: Where it meets the sea near Seward, Alaska, Mears Glacier is more than a mile wide. Note crevasses and pressure ridges. White dots in water are icebergs.

that the ice is hardly visible. The covering of dark material increases heat absorption and accelerates melting, but may reduce evaporation. A very thick mantle of debris may insulate the ice so well that it can last for years in temperatures above freezing.

The bottom end of the glacier, known as the terminus, snout, or toe, is ordinarily a crumbling ice wall, perhaps several yards high, often with an arch or arches through which meltwater from subglacial tunnels comes dashing. This water usually has a milky look because of its content of "rock flour"—the product of grinding by the ice. Below the glacier's snout there is usually a confusion of sand, gravel, boulders, and running water. Here is the dumping ground for rock waste, through which meltwater cuts systems of temporary channels.

Piedmont Glaciers and Ice Caps

At high elevations in temperate latitudes, such as the Alps, and on polar or near-polar highlands such as those of Norway and Alaska, large volumes of glacial ice reach great trunk valleys and the lowlands. Ice issuing from adjacent mountain valleys coalesces to form piedmont ("foot-of-

On Ingraham Glacier, Mount Rainier, Washington: The meeting of many crevasses here has formed seracs. Black lines in seracs indicate yearly accumulations of snow.

the-mountain") glaciers. The motion of these broad masses is very slow, especially at the edges. In the Alps some piedmont glaciers lie largely within the zone of ablation, and their general aspect is one of dirt and decay.

On the high Norwegian mountains and on polar islands such as Spitsbergen, Baffin, and Ellesmere, glacial ice overflows valleys and all but engulfs mountain summits, forming caps over entire regions. As more snow is added, the ice at the edges of caps bulges out and moves downslope through lower valleys to form piedmont glaciers on lowlands or creep into the sea.

On Antarctica the ice cap at its greatest is no less than 14,000 feet thick. Ice at the bottom is believed to be thousands of years old. Radiating from centers of accumulation, the blue rock grinds across the lands, overwhelming or flowing around the mountains. Where it meets one of the high mountain ranges that ring the continent, it divides to flow between the

A piedmont glacier in the Canadian Rockies: Athabasca Glacier, in Jasper National Park, is fed by valley glaciers. Note the ice arch at right.

The Ross Ice Shelf near Little America Five: Here the shelf—actually the edge of an ice sheet—is 100 to 150 feet high. Snow-covered pack ice occupies the foreground.

individual peaks ("nunataks"), forming so-called outlet glaciers, which discharge into the sea. Where no barrier exists, or where a barrier is overridden, the sheet itself pushes out into the water to form an ice shelf.

A glacier entering the ocean is buoyed up because water is heavier than ice. Resulting strains in the ice cause pieces to break off, usually along crevasses. By this process, called calving, icebergs are launched upon their wanderings.

Along some Antarctic shores the ice sheet fills embayments and reaches scores of miles out to sea. The famous Ross Ice Barrier, rising 100 to 160 feet above water, is the edge of one of these seaward extensions. Beyond the ice barriers and the coasts is the zone of pack ice—glacier fragments which, like multitudes of white-sailed ships, drift several hundred miles out to sea.

In Greenland the picture is much the same, though of lesser sweep. The maximum ice depth, as measured by seismic sounding, is about 10,000 feet. This is the depth over the center of the land mass, which like Antarctica has been depressed several thousand feet by the weight of the ice. About half of Greenland's coast is ice-free, particularly the southwestern and northeastern segments, but along some stretches an ice shelf reaches out over the ocean for miles. Much of Greenland's spreading ice encounters mountain barriers and can reach the sea only as outlet glaciers, some of which pour forth ice at rates up to one hundred feet per day.

Sculptured by Glaciers

It is only when glaciers have melted away that the full magnitude of their work on landscapes becomes visible. Their direct work appears in a diversity of rock sculptures and deposits of debris. Other consequences of glaciation include the diversion of streams and the rise of land masses following the melting of ice masses that once bore them down. Signs of glaciation—some imposing and spectacular, others subtle and hardly noticeable—appear on all landscapes that were under ice during the Pleistocene and have since emerged.

Among glaciated mountains a prominent feature is the cirque (in Scotland, corrie; in Wales, cwm; in Germany, kar). This, the mountain hollow where a glacier originated, was given its characteristic amphitheater-like form by frost work and glacial plucking. Meltwater, especially in summer, froze in the joints of the floor and the headwall, or rear wall, of the cirque and bound the mass of glacial ice to the rock. Then, as the ice was squeezed out of the cirque, it plucked out blocks of the rock. This quarrying process can be viewed today on an active glacier when one looks down into the bergschrund—the crevasse between the ice and the headwall.

The headwall is usually steep, and rugged from continual plucking. The sides also are usually steep but may show smoothing due to ice move-

Cirques in the Okanogan Range, northern Washington: A scalloped effect is produced by the close spacing of the cirques. Note the sharp ridges (aretes) between the cirques.

ments. The floor is commonly lower than the threshold over which the ice was discharged, presumably because of greater pressure and greater melting on the floor. The basin thus formed may now hold a cirque lake.

Some cirques were widened until they nearly joined; these are now separated only by the sharp, serrate ridges called arêtes. A series of cirques suggests the edge of a scallop shell. Cirque cutting in mountains on several sides sometimes went far enough to form the distinctive horn peak, of which the Matterhorn is the classic example. Each side of the horn is the headwall of a cirque.

As glacial ice moves down a mountain valley, it crushes, shears, gouges, and abrades the rock surfaces, and breaks off spurs. Part of the resulting debris, together with weathered fragments falling from the valley walls, becomes embedded in the sides and bottom of the glacier, there to be firmly held and used with extreme destructiveness against rock surfaces. Thus equipped, the ice stream can erode its trough much deeper than any stream of water could erode it.

Glaciers rarely make their own troughs; usually they occupy valleys already cut by streams. But glacial ice soon reshapes the original V of the valley into a U cross section, because this offers the minimum resistance to ice flow. The greater the volume of the glacier, the more pronounced the U may be. Accordingly this shape is typical of the great main valleys called glacial troughs, which were once the courses of trunk glaciers fed by many tributaries. Some troughs have been gouged so deep that their floors are lower than the local base level.

Where rivers of water meet, their surfaces usually are accordant—that is, at about the same level. Where a large glacier is joined to a tributary, the top surfaces of the glaciers may be accordant but usually the trough bottoms are not. The larger, heavier ice stream normally has cut its trough deeper. Hence, when the glaciers have melted back, the mouth of the tributary trough is seen "hanging" on the wall of the main trough. Today there is likely to be a waterfall at the mouth of the hanging trough, and beneath the fall perhaps a plunge basin or an alluvial cone or fan.

In highlands where stream erosion has not been controlled by regular, parallel folds, the valleys are likely to zigzag. But where these valleys have been taken over by glaciers, they become considerably straightened. Projecting spurs are lopped off, and after the glacier has melted away, the stumps of the spurs may appear as triangular facets along the valley sides. These facets suggest the ones on eroded fault scarps.

Along valley walls, smoothed and plucked surfaces may alternate. Ice-planed surfaces usually show glacial striations, ranging from hairline scratches to gouges several feet wide, and equally deep. On smoothed bedrock of the valley floor the trail of an ice-embedded boulder may appear as a succession of crescentic marks. Crescentic fractures, up to a foot wide,

Crescentic fractures in sandstone, southeastern New York: Glacial ice moved toward the background. The fractures have survived about 18,000 years of weathering on Schunemunk Mountain.

Crescentic gouges on a granite summit in the Hudson Highlands: The boulder-shod ice moved toward the background, breaking out chips a foot or more wide. Gouges and fractures are preserved on granite in this region but have been weathered off the gneiss.

Glacial striations on sandstone, Bearfort Mountain, New Jersey: These marks were preserved by a soil cover until recently when the land was cleared.

with one- to two-inch spacing, are concave downstream. Crescentic gouges are hollows of about the same width left by the removal of rock chips; they are concave upstream. Lunate fractures are the same except that they are concave downstream. Chatter marks are irregular hollows left by the removal of chips from large grooves.

Bare valley walls sometimes have a borderline above which the rock is much rougher than it is below. This is the trim line, showing the highest level reached by the ice flow. On an overgrown valley wall the trim line may be indicated by an abrupt change in vegetation. Where a glacier has shrunk back during recent centuries, as in Alaska, a forest of mature type, such as spruce and hemlock, may exist above the trim line and a less mature growth, such as aspen and willow, below it.

A valley floor often shows successive rock steps, or glacial stairs, some due apparently to rock variations and others to a sudden increase in the glacier's erosive power where it was joined by a tributary. Near the back of a rock step there is sometimes a basin, usually produced by an icefall. Many such basins are now the sites of ponds.

Larger rock basins were gouged out where heavy volumes of ice debouched into the low-lying glacial troughs. Some of these troughs are now occupied by so-called trough lakes. Where trough ice bulldozed onto the lowland, it sometimes dug there an even larger basin before spreading out to form a piedmont glacier. These large basins now may hold piedmont lakes.

Strongly glaciated valleys along coasts have become fiords. A fiord is the seaward segment of a valley gouged below sea level by glacial ice and later filled by the post-Pleistocene rise in sea level. Fiords have become partly filled with recent river sediment, but borings find bedrock hundreds of feet beneath it. At the point where the ice entered the ocean and was buoyed up, downcutting ceased, and a rock threshold now marks the spot. At the threshold there is usually also a deposit of bulldozed rock debris—a terminal moraine.

Breathtakingly scenic fiords indent the coasts of Norway, Greenland, Alaska, British Columbia, Chile, and New Zealand. Specimens are known also along the coasts of Yugoslavia and Scotland. Norway's Sogne Fiord, one of the greatest, leads the sea 120 miles inland; its bottom at deepest lies 3700 feet below sea level, and its walls rise several thousand feet above water. Messier Channel in Chile has a depth of 3900 feet, Finlayson Channel in British Columbia 2400 feet, and the outer part of Chatham Strait in Alaska 2700 feet.

Some valleys that are not generally recognized as fiords really are. Among them is the lower Hudson Valley of New York state, which in the Hudson Highlands is a glacial trough. Drillings through the bottom sediments here have reached bedrock at nearly 900 feet below sea level,

ROCK, TIME, AND LANDFORMS · 234

whereas downstream the distance to bedrock does not exceed 275 feet.

Horn peaks, cirques, U-shaped troughs, and hanging valleys are all typical of what is called alpine glaciation—that is, glaciation on high mountains. These features are being shaped even today on terrains such as the higher Alps and the Himalayas. Alpine sculptures of Pleistocene origin are common in the mountains of Norway, Scotland, New Zealand, and southern South America. In North America strong sculpturing occurred as far south as the Sierra Nevada and the Colorado Rockies. Yosemite Valley, with its famous waterfalls from hanging valleys, is one of the world's

The great glacial trough of Banff National Park, Alberta: During the Pleistocene ice filled this valley, grinding it into the characteristic U-shape.

ICE ON THE LAND · 235

greatest troughs, the Teton Range in Wyoming is especially rich in horn peaks, and the Colorado Rockies are memorable for great cirques and chains of rock-basin lakes. The northern Rockies from Montana to Alaska are unsurpassed as a showplace of glacial carving.

A different kind of ice work was done during the Pleistocene on lesser highlands in the temperate regions. Here the ice sheets rode up over the hills and overwhelmed them. Valleys beneath the sheets were smoothed but not deeply gouged, so that today the highlands usually have rounded, sweeping profiles. The upstream slopes of many have been planed, and the

downstream slopes plucked. Similarly planed and plucked slopes are seen on knobs of resistant rock (sometimes called *roches moutonnées* because they slightly resemble old-fashioned wigs slicked down with mutton tallow). Around the summits one may come upon potholes formed by meltwater streams that cascaded down through moulins and crevasses while the summits were still covered by the ice. Along some main valleys that are aligned with the general ice movement truncated spurs may be seen. The brook beds are still loaded with glacial rubbish of all sizes, as are ravines in which today there are no streams at all.

Ice-smoothed highlands are familiar northward from Minnesota, the northern portions of Pennsylvania and New Jersey, and New York and New England. Northern England, Ireland, Wales, and southern Scotland also have them. The northern sectors, as in the White Mountains of New Hampshire and the Catskills and Adirondacks of New York, show some well-developed cirques. More common is the nivation hollow, formed where snow accumulation was not quite sufficient to make a glacier and a cirque. Valleys in line with the ice movement show something of the U-shape, though many have become partly filled with recent alluvial deposits. Some

A New Zealand fiord: Milford Sound is one of the most spectacular of drowned glaciated valleys. The beach in the foreground is formed of glacial outwash.

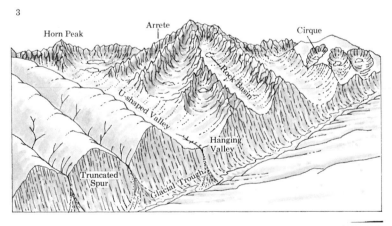

Erosion by valley glaciers: The ice alters the details of scenery, sharpening the profiles, but the general pattern of mountains and valleys remains.

A roche moutonnée: The ice movement was from left (north) to right. Note the results of plucking on the south side. The locality is the Hudson Highlands.

of the upstream slopes, especially those covered with protecting soil for part of the time since the Pleistocene, still carry the signatures of glacial ice—chatter marks, crescentic fractures, grooves.

The ice sheets left their marks also on the bedrock of lowlands. In central and eastern Canada, where much of the land was stripped of its soil by southward-moving ice, the bedrock was intensively polished, scratched, grooved, and gouged. Seen from the air, some of these landscapes appear to have been clawed by giants. Long, straight grooves in the rock parallel low ridges of moraine dragged out by the ice. Where zones of weak rock were deeply gouged, and where basins were shaped in the debris, there are today thousands upon thousands of lakes.

Lowlands in the extreme northern United States, particularly in Maine, Wisconsin, and Minnesota, were likewise scoured. These lands also are patched with hosts of rock-basin and other lakes. On most of these areas, however, the final melting of the ice covered the bedrock with debris on which forests and meadows could develop.

Deposited by Glaciers

The erosive activities of glaciers have produced enormous volumes of rock waste. Wastes from valley glaciers have been carried down by the ice and by streams into trunk valleys and out onto the lowlands. The wastes borne by the Pleistocene ice sheets and dropped where they melted lie scattered on highlands and lowlands alike over more than a third of the planet's land surface.

Mountain valleys from which the glacial ice has vanished show evidence of torrential streams in the past. Potholes and extensive water-worn rock pavements are found where only little streams trickle today. Valleys are littered with stream-borne debris, including boulders yards in diameter. Some of this material was rounded by abrasion beneath the glacial ice and by tumbling in the meltwater; some is still angular despite thousands of years of postglacial weathering.

Near the foot of deeply glaciated valleys there are often lateral moraines, consisting of debris carried along at the edge of a glacier and left by the melting. The farthest point reached by the tip of the glacier may be marked by a terminal moraine—a heap of unsorted rock waste suggesting the work of a gigantic bulldozer.

Some debris accumulations were well sorted and stratified by meltwater streams. Among them is the esker, a long and sinuous ridge of

A glaciated landscape in northern Canada: Grooving and basin-making by Pleistocene glaciers are striking. Grooves and elongation of the lakes indicate direction ice moved.

Eskers in the Punkaharju region, Finland: Streams of glacial meltwater left deposits in the form of these sinuous ridges. The region is a mosaic of lakes.

sediments left by a stream that flowed over, within, or beneath a glacier. Having been superimposed on the land by melting, an esker may run down into hollows and up over knolls regardless of the changes in elevation. Valley trains, also sorted by meltwater, are masses of glacial debris, usually coarse, extending downslope from a terminal moraine and covering the valley floor. Beyond the valley mouth, sorted debris often fans out to form an outwash plain.

Glaciated lowlands may carry a wide mantle of ground moraine: unsorted material dropped randomly by the melting of a broad ice mass. Where meltwater poured over the edges of such a mass it sometimes formed hummocky ridges—lateral moraines—up to one hundred feet high. Another kind of meltwater deposit, the usually domelike hill called a kame, formed where debris-loaded water cascaded over the side of a glacier, or into a crevasse or a melthole (moulin), or down a slope against the ice. A lateral moraine containing many of these small hills is called a kame terrace; it occurs where a stream flowed along the edge of an ice sheet.

Ice riding over moraine sometimes shaped it into a drumlin. This

Glacial drift cut by gullying near Madison, Wisconsin: The material was not sorted by size, and hence probably was not deposited by a stream.

usually happened where the ice met a rock knob or other obstacle and molded rock waste around it. Drumlins are elongated hills as much as five hundred feet high, occurring in clusters and sometimes attaining lengths of many miles, the long axis being in the direction of ice flow. Seen from the side, a drumlin has a steep slope on the upstream end and a gradual slope on the other. Wisconsin, eastern Massachusetts, and central New York are rich in drumlins, as are the Scandinavian countries. Perhaps the most famous drumlin is Bunker Hill in Boston.

Deep, broad moraine deposits usually contain depressions called kettles, made by the melting of ice blocks buried in the moraine by stream action or by being overridden during a glacial advance. As a block melted, the covering material sank down over it to form a kettle. Today lakes, ponds, and swamps occupy kettles whose bottoms are below the water table. Kettles are commonly associated with kames, forming hummocky relief of the kind seen on Cape Cod and Long Island.

During melting at the close of the Pleistocene, glacial debris accumulating in valleys ponded many streams. Most of the natural dams have

A kame in Sussex County, New Jersey: Such heaps of glacial debris, left by meltwater pouring off an ice sheet, today offer unlimited supplies of cobbles, gravel, and sand.

since been breached by overflow, but the debris remains. Some of the empty basins are now broad plazas of alluvium crossed by little brooks.

Till and stratified drift—clay, sand, gravel—lie thick on glaciated flatlands. In the central United States, particularly Ohio, Indiana, and Illinois, the thickness of till averages one hundred feet, and the thickness is much greater where preglacial valleys were filled. During the great meltings the landscape was laced with meltwater streams which spread the rock waste to make an almost perfectly level surface. But for the ice sheets, this region would be today a land not of plains but of gentle hills and valleys.

Analyses of Midwestern till indicate that some must have come from sources hundreds of miles to the north. Erratics found in Missouri have been traced to rock formations in southern Canada. Likewise, boulders on Long Island have been identified with rocks of New Hampshire, and erratics in northern New Jersey have been derived from central and northern New York. As mentioned earlier, an estimated thirty to forty feet of the rock and soil of central and eastern Canada was carried by the ice sheets down into the United States.

Sheet-glaciated highlands also show remarkable glacial littering. On broad summits there are many boulders, the lighter drift having been mostly washed or blown to lower levels. In ravines and valleys the glacial litter mingles with that of the fierce frost-wedging of the Pleistocene and thousands of subsequent winters. It clogs the streambeds—both those where streams run today and those that have been streamless for millennia. Some streams have been diverted by the drift, and damming has produced chains of ponds and swamps.

On lowlands, too, stream diversion was a common result of glacial invasions. Channels well established before the Ice Age were in some instances obliterated as the ice ground over them. After the ice melted, some

Erratics on an ice-planed summit, Hudson Highlands: Most of this rock litter comes from hills immediately to the north. Some has been transported scores of miles.

In a boulder-filled valley, Hudson Highlands: This small upland valley was a spillway for glacial meltwater. The thick glacial debris makes the area all but impassable.

Loess in a road cut west of St. Louis, Missouri: The yellowish material slices cleanly and stands with almost rocklike firmness despite slopewash from rains.

streams resumed business in old channels, but others were deranged by the accumulations of rock waste. In the Midwest particularly drainage changed radically as meltwater filled old valleys with rock waste and cut new ones. Some preglacial rivers, such as the Teays, which flowed northwest over Ohio, Indiana, and Illinois, did not survive the glacial invasions. The present Ohio River, which originated along the edge of an ice sheet, occupies not only the valleys of the preglacial Ohio, west of the present site of Cincinnati, but also part of the old Teays Valley and part of the former Allegheny-Monongahela drainage system. The Missouri likewise was deranged by the Ice Age, as were the upper reaches of the Mississippi.

A landscape glaciated by an ice sheet: The features are mostly depositional. Generally they are typical of those resulting from glaciation of a terrain of low relief.

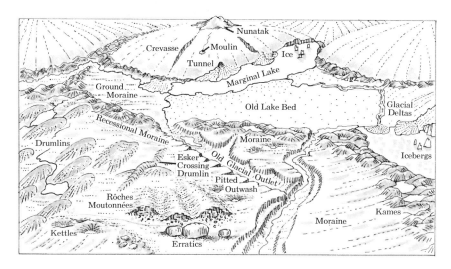

An elevated coast: The ice-planed land of North Cape, Norway, has gradually risen since its burden melted away. Most other northern lands also are rebounding.

Glacial rock wastes include the curious material called loess. Consisting mostly of silt-sized particles, predominantly quartz, and in color ranging from yellow and orange to brown, loess is found in wind-deposited layers on prairies and deserts and on terrains nearby. Some loess has come from unglaciated deserts, but most of it was swept up from glaciated lands by the wind after the Pleistocene, before vegetation could anchor it.

Loess compacts well and can stand as cliffs where streams and roads cut through it. It may be given a columnar appearance by systems of vertical joints, some of which follow the courses of plant roots. Loess cliffs are seen along the courses of the Missouri and Mississippi rivers and their tributaries, notably in Illinois and Mississippi. There are 40-foot deposits in Kansas. In China loess deposits reach depths of 1000 feet, and peasants formerly dug into them to make cave dwellings. In Eurasia loess is abundant from France eastward all the way to China, and in South America it lies on the plains of Argentina and Uruguay east of the Andes. Loess carried by winds from Central Asia lies thick among the upper sediments of the western and central Pacific Ocean bottoms.

Different kinds of glacial deposits predominate in different regions. Outwash plains cover large areas of northern Europe, especially Denmark and Sweden, and in the United States are seen mostly in the north-central states. The upper part of the Mississippi Valley is an outwash plain. Eskers are most common in Finland, though many can be counted in Denmark and Sweden, along with kames, moraines, and drumlins. In North America long eskers wind across the Canadian plains, and some are found in the upper Mississippi Valley and in Maine, New York, and northern New Jersey. These regions also have many moraine features, as does Cape Cod —which, incidentally, consists entirely of sand and clay deposited at the edges of an ice sheet. Wisconsin, eastern Massachusetts, and New York are rich in drumlins.

Lands Down and Up

Centuries ago, people who lived along the margins of Europe's Baltic Sea noticed and pondered the shell-strewn terraces hundreds of feet above sea level. Shore-dwellers of Scotland and of England's North Sea coast likewise must have wondered about what were evidently elevated beaches. Along North America's Great Lakes, settlers noticed raised beaches which clearly represented former high levels of the lakes. To the Europeans it must have seemed that the sea was falling; to the Americans it must have seemed that the lakes were gradually drying up. But with our modern knowledge of the Ice Age, these abandoned shorelines are easier to understand. Pleistocene ice depressed the crust, and now the crust is rebounding.

Ice weighs about a third as much as average rock. When glacial ice reaches a depth of some hundreds of feet, the crust begins bending beneath it. Rock at depth gradually flows outward from the center of deepest depression. As the pressure is relieved by ice decay, the flow is reversed and the crust over a period of perhaps ten thousand to twenty thousand years recovers approximately its former profile. In areas where Pleistocene ice was thousands of feet thick, lands are still rising today.

Central Scandinavia, the site of an ice sheet probably 8000 feet thick at maximum, has risen about 700 feet during the past 9700 years. The present rate of rise, as indicated by long-period tide records and beach chronology, is about 38 inches per century, but it is slackening. The rate of uplift and the total uplift since the Pleistocene decrease with distance from the center of the ice cap, which was about 100 miles west of Namsos, Norway. Uplift apparently has ceased south and southeast of the Baltic Sea, but it is comparatively rapid near the junction of the Baltic with the Gulf of Bothnia, and these two bodies of water may become separated a few thousand years hence, so that the Gulf will become a shrinking freshwater lake.

The rate of uplift in the Scottish Highlands, the site of another former ice cap, is now about sixteen inches per century. In the British Isles, however, the picture is complicated by the fact that rebound is being offset, at least in the southern region, by a submergence apparently unrelated to glaciation. Uplift is still in process over a roughly oval area that includes Scotland, Northern Ireland, and the northern parts of England and Eire.

Surveys of the Great Lakes reveal that the old shorelines are higher on the north sides than on the south. In the north the land was depressed more, and hence has had more elevation to recover. Since the ice to the north melted away more recently, rebound there is more rapid. Northward from a line connecting central Michigan, Lake Erie, and southern Maine, the rate of uplift increases from a value of zero to about twenty inches per century, the latter being the value along a curve extending from northern

Lake Superior to Lake Ontario and northeastward along the Gulf of the St. Lawrence.

The Hudson Bay region was the center of one of the greatest North American ice caps, and here uplift has been going on for an estimated 5000 years—presumably since the ice cap thinned to a depth of 1000 feet or so. Since recent marine sediments have been found here at altitudes of over 900 feet, and since the post-Pleistocene sea-level rise has been about 275 feet, 1175 feet is a rough minimum value for the recovery of this region to date. Future uplift is expected to total at least 1000 feet—enough, perhaps, to empty Hudson Bay.

The Shrinking World of Ice

Melting after the last glacial maximum was well started by 15,000 years ago, but within historic times several notable climatic fluctuations have occurred. A period of accelerated warming came between about 4000 and 6000 B.P. (before the present). This period was followed by a cooler one. Another warm interval, during the Middle Ages, was followed by glacial advances—some marked by the razing of Alpine towns—late in the

A severely shrunken glacier on Mount Rainier: Emmons Glacier shrank greatly after the Pleistocene but has been growing in recent years. Note the heavy accumulation of moraine in the valley.

sixteenth century and early in the seventeenth. A warming trend which began late in the nineteenth century is still with us.

Mountain glaciers of temperate regions today include some impressive specimens. Fedtchenko Glacier, in the Pamir Mountains north of India, is 48 miles long and as much as 3000 feet deep, and it boasts no less than 25 tributaries. In the towering, ice-mantled St. Elias Mountains of Alaska, Hubbard Glacier has a length of nearly 75 miles—perhaps the greatest on Earth. Malaspina, a piedmont glacier in Alaska's Chugach Range, attains a thickness of 1500 feet and a width of 40 miles, and it traverses a 28-mile-wide coastal plain on its way to the sea. The Alps, with about 1000 glaciers in all, exhibit many with thicknesses of more than 2000 feet and lengths up to 10 miles. Even Kilimanjaro and several other lofty African peaks have glaciers of respectable bulk.

But current rates of melting are high, and most of the world's glaciers are shrinking. Only a few, such as those on Mount Rainier in Washing-

Panorama of alpine glaciation: Horn peaks, cirques, serrate ridges, and troughs are abundant in this view. The Emmen Valley is in the foreground. In the far background are the Wetterhorn (*extreme left*) and the Jungfrau (*extreme right*).

ton, are—because of special local conditions—growing. The Rhône Glacier in the Alps is typical of many in having lost no less than 3 miles of its length in the past century, and Alaska's Muir Glacier during recent decades has been retreating about 1500 feet per year. In Alaska, where shrinking has been especially rapid, a number of glaciers have been shortening at rates exceeding 500 feet per decade. In Scandinavia and the Alps the snow line has risen 300 feet during the past 50 years, and hundreds of small glaciers have dwindled away. During the same period, glacial meltings have raised sea level about 2 inches, and naturalists have observed a steady northward advance of wildlife habitats.

The reason for glacial shrinkage is well known: during the past fifty years average temperatures over the planet have increased by 2 degrees to 10 degrees Fahrenheit.

According to scientific records, glacial recession speeds up during periods of maximum sunspot activity. These are periods of maximum insolation—that is, of maximum radiation received by Earth from the sun. It is thought, therefore, that glacial ages, which have occurred at about 250-million-year intervals, may be due to long-period variations in insolation. Reductions in insolation could result from decreases in the sun's energy output and from increases in the shielding effects of Earth's atmosphere. Such increases can occur when volcanoes hurl extraordinary amounts of dust into the stratosphere.

Since glacial ice forms mostly on high mountains, it presumably increases with mountain-building activity. After periods of rapid uplift (from one of which we may now be emerging) glaciers should wane as highlands are reduced by erosion. It has been suggested too that since carbon dioxide in the atmosphere traps solar heat, the rapid output of this gas in our age of technology may be causing a planetwide warming. As with many other problems of geology, the solutions will be discovered only by decades of further scientific observation.

XI

Lakes and Swamps

A rock-basin lake in the Adirondacks: The depressions occupied by Lake Placid are believed to be fracture zones gouged out by the ice sheet. Rectangular fault patterns are common in this region.

Lakes, which rank among the most popular kinds of scenery, are also among the shortest-lived of all. Only an extraordinary geologic event can bring a natural lake into being, and then its built-in suicidal features combine to eliminate it.

Unusual geologic events have put more lakes on the planet in the modern era than during most of the past. As the Pleistocene ice bulldozed our landscapes and plastered them with billions of tons of rock waste, it created tens of thousands of natural basins that are today filled with water —not the meltwater of the glaciers but the drainage of today. And besides these ice-made basins there are many others. Our era follows a peak of diastrophic and volcanic activity. River valleys have been dammed by landslides and lava flows; grabens have been formed by faulting, and craters by volcanic activity. Even after subtracting from our lake total all those made by man's dams, it can be said that rarely if ever in the geologic past has there been a period of lake-making to match the Pleistocene.

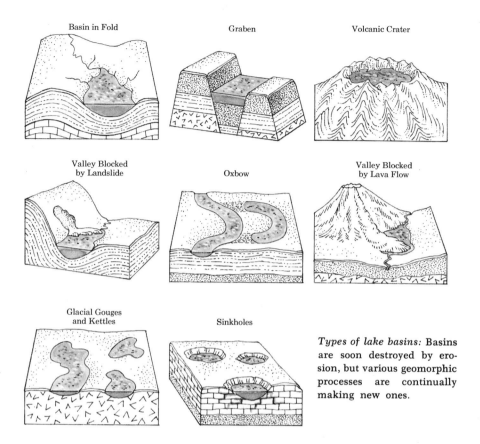

Types of lake basins: Basins are soon destroyed by erosion, but various geomorphic processes are continually making new ones.

A Lake's Basin

A natural lake is essentially a body of water in a basin formed by rock or soil. Its surface, like that of a river, usually approximates the water table of the vicinity. Water is in the basin by virtue of its tendency to settle to the lowest possible elevation. The basin normally is not watertight; ordinarily some ground water circulates through bottom and sides. A lake is supplied also by direct precipitation and runoff. All lakes have one or more inlets, consisting of surface streams or springs, and most of them have outlets—usually at the surface—through which water is discharged. Many lakes occupy basins which, by some accident of nature, have been created in the course of a stream.

For some lakes, such as those on flood plains, ground water is the main source of supply. The basins lie in porous soil through which water percolates easily. Many lakes lie in bedrock or close-packed soils through which water moves slowly, so that they may depend more on runoff than on ground water. Here the water level fluctuates widely according to current rainfall.

Some rock-basin lakes get ground water through natural pipelines formed by joint systems in the bedrock. A few, like artesian wells, are fed by hydrostatic pressure; that is, by water descending from higher elevations between impervious strata. The surfaces of these lakes can be higher than the water table in land immediately around them.

Most lakes that are not supplied by ground water are temporary. Typical of these are the playa lakes of the desert and mountain ponds that are deep with meltwater in spring but dry by midsummer.

Many swamps are simply lakes that are shallow enough so that plants growing from the bottom can reach the surface. A high proportion of swamps are, in fact, lakes that have become almost filled by vegetation. Some swamps are lakes that have been lowered by downcutting of the outlets, and in others the dearth of water is due to a dry trend in climate. In swamps generally, plants are taking over and, barring interference, will eventually convert the sites into dry land.

Swamps are common on valley bottoms and coastal plains where gradients are low, the ground porous, and the water table just below the surface. Also, prairies and deserts have natural catchment basins, usually in nearly impervious bedrock, where rainwater is held for weeks or months —long enough to support vigorous vegetation—although the water table may be far below the surface. The plants shade the water, retard the air circulation, and so reduce evaporation. In humid regions swamps can develop on hillsides where tightly knit plant roots slow the descent of water from springs or melting ice and snow.

Since their basins are shallow, swamps respond more obviously than lakes to changes in weather. A few days of rain may convert a swamp into a lake; a few weeks of drought may transform it into dry ground. Many swamps are, in fact, just temporary features that we owe to fall rains or spring melting.

A World of Glacial Lakes

An air traveler over northern lands, particularly above 45 degrees north latitude in North America and 55 degrees in Eurasia, is entranced by the vast panoramas of lakes. North America sparkles with lakes especially between Maine and Minnesota and northward throughout Canada. Lesser but still impressive numbers of lakes dot our landscapes as far south as New Jersey and Colorado.

In Europe lakes are plentiful from northern Germany into Scandinavia, and in mountain terrains to the south. In Asia the lake region reaches from the central areas of the Soviet Union northward to the Arctic.

These scattered gems of nature show wide diversity in shape and size, ranging from inland seas such as America's Great Lakes to ponds and swamps too small to show on a county map. Virtually all are relics of the recent glacial ages; few preglacial lake basins could have survived the scouring by the ice sheets. More than two thirds of all today's natural lakes are of Pleistocene origin.

Common among high mountains that were sculptured by valley glaciers is the lovely cirque lake, which is made possible by the threshold that typically exists at the outer edge of a cirque. Rock flour gives many of these lakes a turquoise hue, which is strongest when a blue sky is being reflected.

Descending a glacial valley from the cirque, a walker may find a series of water-filled hollows on rock steps along the valley bottom. These lakes are most commonly found in the upper valley, often at sites where the entrance of a tributary glacier caused deeper gouging. They are called

A cirque lake in the Teton Range, Wyoming: The basin for Lake Solitude is the floor of a cirque from which glacial ice once moved down through the north fork of Cascade Canyon. At upper right is the Grand Teton.

Paternoster lakes in the Rockies: Lakes Grinnell *(foreground)*, Josephine, and Sherburne in Glacier National Park occupy basins gouged by Grinnell Glacier.

paternoster lakes because, being in series and connected by a stream, they suggest beads on a rosary.

Cirque and paternoster lakes sparkle on higher ranges of western North America as far south as the Sierra Nevada. They occur in Europe all the way down to the southern Alps, and there are many in Scotland and in England's Lake District. A few cirque ponds can be identified in the White Mountains of New Hampshire and the Adirondacks of New York.

Near the lower ends of trunk valleys, where they level out before reaching the lowland, so-called trough lakes appear. Such lakes are massive —Lake MacDonald in Glacier National Park, Loch Ness in Scotland, and Lakes Como, Lugano, and Maggiore in Italy are examples. Usually the basins have been dammed at the lower ends by end moraines.

Where glacial ice comes down onto a lowland, it may dig deep before spreading out to melt and die. During the Pleistocene, adjacent valley

The Great Glen in Scotland: This ancient river valley, dug deeper by Pleistocene glaciers, is now occupied by lochs. Loch Ness and Loch Oich are seen here.

glaciers emerging from their valleys to form piedmont glaciers dug the basins for our piedmont lakes of today. A famous one is Lake Geneva in Switzerland.

The gouging capabilities of ice sheets are evidenced by the multitude of lake basins on glaciated flatlands of northern Eurasia and North America. Many such basins were scooped out entirely by ice work; generally they are elongated in the direction of ice movement. Numerous river valleys were reamed out and then dammed by erosion debris. Where the valleys were aligned approximately with the ice trend, very deep basins such as those of Lake Michigan and New York's Finger Lakes sometimes resulted.

Glacial basins include many that were shaped entirely in glacial moraine—the litter of sand, clay, gravel, cobbles, and boulders. Kettles, the depressions formed where buried ice blocks melted, account for many basins. Other morainal lakes occupy basins where sand and clay were bulldozed into heaps and ridges by the ice and further shaped by meltwater streams. A high proportion of these have dried up since the Pleistocene. Where compacted clay underlies the hollows, however, as on Cape Cod, drainage is slower, the water table higher, and more lakes survive.

LAKES AND SWAMPS · 257

Ten to twelve thousand years ago, when the last ice sheet began to melt, the lowlands of central North America and northern Eurasia were half-drowned. Terrain at the edges of the ice, downwarped by the weight, collected much water as so-called ice-marginal lakes. These expanded as the ice melted back toward the north. The greatest of these bodies of water in North America was Lake Agassiz, which at maximum extended eight hundred miles northward from South Dakota and covered parts of North Dakota, Minnesota, Saskatchewan, and Manitoba, drowning more land than all of today's Great Lakes together. Ice-marginal lakes were features of late-Pleistocene Scotland and England also. Today they form in Norway, Iceland, and Greenland at the edges of glaciers on the lowlands.

The glacial lakes of today are few and shrunken compared to those of yesteryear. But their ghosts still lie around us. The swamps of northern lands are mostly glacial lakes that are nearly filled with vegetation and rock waste. Flying over the woodlands of Maine, northern Wisconsin and Minnesota, and central Canada, one sees these ghosts in uncounted multitudes —brownish patches of all shapes and sizes in the dark seas of evergreen.

Lakes in Structural Basins

The perennial disturbances of Earth's crust might be expected to produce in every age a host of structural basins. Actually, such basins are

The English Lake Country: Like other bodies of water in this famous region, Lake Derwentwater owes its basin to the Pleistocene glaciation.

Kettle lakes on Cape Cod: Gull Pond (*center*) and Higgins Pond, near Wellfleet, occupy depressions that formed over buried Pleistocene ice blocks as these melted away.

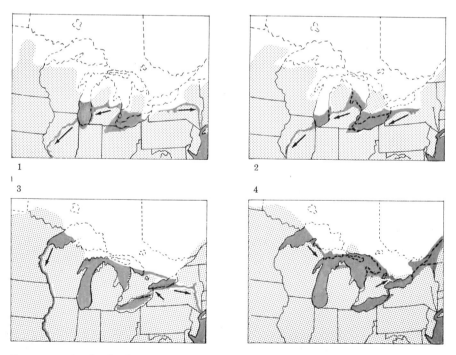

Four stages in the development of the Great Lakes: Arrows show how the direction of drainage changed as the front of the Pleistocene ice sheet advanced and melted back.

A graben lake: North-south fault scarps bound Lake Tahoe on the east and west. In the distance is the Nevada side. The lake is over 1600 feet deep.

relatively rare. Folding and faulting do not commonly produce the basin form, and where they do erosion is usually rapid enough to maintain an outlet at one side of the depression so that it cannot hold water.

Some notable structural lake basins do, nevertheless, exist. Lake Superior's great depth is due partly to downwarping of its bottom under Pleistocene ice. In Europe the basins of some mountain lakes, such as Maggiore and Lucerne, are in downfolds deepened by glacial gouging and dammed by moraine. North America's Basin and Range Province has the downfaulted depressions called bolsons, which are sites of temporary lakes.

The coastal plains have some lake basins that geologists attribute at least partly to crustal warping. A number of these basins communicate with the sea through channels that are kept open by tidal currents, and hence the lakes are properly regarded as bays or lagoons; but where basins have become separated from the sea by uplift, warping, or the formation of wave- or wind-built barriers, true lakes exist. In these the water is fresh, brackish, or salt according to whether the dominant source of supply is runoff, fresh ground water from landward, or salt ground water from seaward. North America's best-known coastal-plain lake is Florida's freshwater Lake Okeechobee, which is some 750 square miles in area and has a maximum depth of 15 feet, with a water level averaging only 20 feet above the level of the sea.

Most diastrophic lake basins are grabens, and some are very deep. Lake Tahoe, on the border of California and Nevada, is 1654 feet deep, and Lake Baikal in Siberia is 4500 feet in depth—the deepest lake in the world. This lake's bottom is more than a half mile below sea level—clear evidence that the basin was produced by faulting, not by erosion. Lakes Saranac and George in New York state also are graben lakes.

During the pluvial, or wet, conditions that prevailed during the Pleistocene meltings, numerous and vast lakes patched the Basin and Range Province and similar lands of poor drainage in North Africa, the Near East, and Central Asia. In the Basin and Range Province the greatest of these lakes was Bonneville, which at maximum was 1000 feet deep and covered over 20,000 square miles in parts of what is now Utah, Idaho, and Nevada. Another gigantic lake was Lahontan, which drowned about 8000 square miles of Nevada, California, and Oregon to a depth of at least 500 feet.

Fault zones, which often become important drainage lines, sometimes are blocked by new faulting episodes. Along the San Andreas Fault in California are narrow valleys with small, water-filled basins formed by sags and ridges. Where major rivers flow through fault zones, they are ponded intermittently by the rise of new scarps.

Numerous lakes occupy grabens along Africa's Great Rift, among them the Dead Sea, the Sea of Galilee, the Red Sea, and Lakes Tanganyika and Nyasa. The top of the sunken block on which the Dead Sea lies is a half mile below the surface of the Mediterranean, and the surface of this graben lake is itself 1300 feet below sea level.

River-made Lakes

Ponds and sizable lakes occupy some of the plunge basins of rivers, particularly those that were much more voluminous in the past than they are today. Plunge basins are the hollows ground into bedrock beneath waterfalls. In the United States one of the best known is the basin shaped by the Columbia River near Grand Coulee in Washington while this river was carrying huge amounts of meltwater from a melting ice sheet immediately to the north.

On flood plains such as that of the lower Mississippi there are the familiar oxbow lakes. It will be recalled that these are created at times of high water when the river, flowing faster, cuts off meander loops. Later, as the torrent subsides, sediments seal the ends of the loops, separating them from the river. Some oxbow lakes last only until the next flood, at which time the rushing waters obliterate the old drainage lines and make new ones. Other oxbows survive but, being typically shallow, tend to fill with vegetation, become swamps, and eventually dry up. Filled oxbows are features not only of today's flood plains but also of regions once overrun by glacial meltwater.

On deltas, too, there are lakes, lying in basins shaped by the rivers' distributaries. As floods subside, the distributaries drop sediments along their own courses and, sometimes, athwart the courses of other streams. The waters of the blocked streams then back up, forming lakes. Like oxbows, delta basins are shallow and extremely vulnerable to flooding, and thus are short-lived. Among the few large ones that can be seen today is Lake Pontchartrain, on the Mississippi Delta, which has maximum dimensions of about 20 by 40 miles, its depth being around 27 feet. The Zuider Zee, north of the Netherlands, lies in a vast basin shaped by distributaries of the Rhine Delta.

Blocked-Valley Lakes

A so-called fan lake originates where a stream is dammed by the delta of a tributary. Tulare Lake, in California, was created where the King River, descending from the Sierra Nevada, dumped loads of sedi-

Oxbow lakes: A high-altitude view of the Yukon Flats in Alaska, where the Yukon River has meandered for millennia. Note the numerous cut-offs and meander scars.

ments to make the large fan that blocks the San Joaquin River. The lake has accumulated upstream from the fan.

In Switzerland a stream descending northward from the Jungfrau has built a fan across an old lake to form the modern twin lakes Thun and Brienz. The town of Interlaken stands on the fan.

During recent geologic time great amounts of lava dust have accumulated on the volcanic San Juan Mountains of southern Colorado. About a century ago a thick mass of this material became saturated from heavy rains. The muck oozed down the mountain slopes, filled a valley, followed it for four or five miles, and finally spread as a broad, fanlike barrier two hundred to three hundred feet high across the canyon of the Lake Fork of the Gunnison River. Thus dammed, the river waters backed up three miles before an outlet was found at the edge of the mudflow. The body of water thus created is Colorado's famed Lake San Cristobal.

In 1925, in the Gros Ventre Mountains of Wyoming, some 50 million cubic yards of soft, rain-saturated shale and sandstone slid down into the valley of the Gros Ventre River. Raising higher the natural dam formed by previous slides, it converted Slide Lake—a mere pond—into a body of water three miles long. Two years later the dam broke, and most of the stored water rushed downvalley, devastating the small town of Kelly. A reduced but still substantial Slide Lake survives.

In desert country mudflows frequently dam streams to form lakes that are only temporary. In mountain terrain under a humid climate landsliding is frequent, and when masses of rock, soil, and mangled trees block a ravine, a pond or small lake can soon form. A high proportion of mountain lakes are of such origin. Dams made by slides are, however, more quickly cut away than are the edges of rock basins, and most slide lakes thus have a short life expectancy.

Valley blocking can occur by volcanic action also. Snag Lake, in Lassen Volcanic National Park, California, is but one of countless bodies of water in volcanic regions that have been created by damming of a stream by a lava flow. A large valley in the Great Rift of eastern Africa has become blocked in recent time by the rise of the volcanic Mfumbiro Mountains, and the result is Lake Kivu. On a smaller scale are the ponds in places like Yellowstone National Park where basins have been formed by accumulations of minerals from hot springs and geysers.

Basins Made by Plants and Animals

Some natural dams are the works of plants and animals. The engineering projects of beavers, which have long interested and sometimes frustrated man, were undoubtedly of importance to our landscapes before man's arrival. Plants, on the other hand, work more slowly and subtly,

Dammed by a landslide: Earthquake Lake in southwestern Montana originated in 1959 when a great slide in the Madison Range blocked the Madison River.

being most effective where streams are shallow and slow. Swamps rather than lakes are the typical results of their work.

A notable basin created by plant growth is Lake Drummond, in the Great Dismal Swamp of Virginia and North Carolina. Even after being reduced by artificial drainage, the swamp covers about one thousand five hundred square miles and the lake, about five miles in diameter, is six to ten feet deep. The waters are contained by thick parapets of peaty material accumulated over thousands of years.

Plant-built basins are common in the tundra country of northern Canada, Alaska, and Siberia. Here drainage is poor because of the flatness of the land and, in many localities, the presence of impervious subsoils. The subsurface is permanently frozen, and downward percolation of water is slight. During the summer water accumulates in broad, shallow ponds and swamps. In the South Pacific some coral islands contain basins which originally formed as lagoons, then were raised above sea level by the uplift of the islands, and now hold fresh water.

In the warmer localities of Africa and South America, large animals roll in the cool mud at springs and in depressions where poor drainage has allowed water to stand after rain. Hooves break up the rock, bit by bit. Later the winds blow away the loosened rock particles to make basins. In

Africa some waterholes have eventually become ponds by this process.

Among animal lake-builders man also must be listed. Most landscapes contain river valleys that can be dammed at small expense, and when plenty of money is available veritable inland seas can be created by means of giant dams. In parts of the Midwest ponds can be made simply by excavating a few yards down into glacial sand and gravel to a level a little below the water table. Even where the water table is too far down to reach without excessive expense, ponds can sometimes be maintained temporarily on shale or other impervious strata.

Uprooted trees and other plant debris, swept along by a river in flood, often accumulate to form rafts. Such a raft sometimes plugs a water course so effectively that a lake builds up behind it. The most famous timber-raft dam in United States history was the one that started building in the Red River, in Louisiana below Alexandria, some time before the year 1600. A leaky dam at best, but nevertheless functional, it kept growing until 1873, at the rate of a mile per year, creating lakes upstream as it grew. While new material accumulated on the north end, the old material at the south end decayed and was carried downstream; thus the great raft traveled upstream in a state of continuous decay and creation. It was broken up finally, in 1873, by rivermen working from steamers with cranes and saws. Most of the lakes that it made have since been drained, and their bottoms are now the sites of flourishing farms.

Basins by Solution

Regions with limestone bedrock at the surface are likely to be dotted with many small natural rock basins, many of them containing water. These basins result from solution of the limestone. The so-called solution

A sinkhole pond in Virginia: The pond occupies a broad depression in the limestone near the entrance to Luray Caverns. Note the lack of any surface outlet.

A lake in a volcanic crater: Crater Lake, Oregon, occupies the caldera of the extinct volcano Mount Mazama. The crater was active until about 6300 years ago.

hollows, some of them square miles in area, but usually very shallow, form where the rock is weakest or is less protected by vegetation. Here and there funnel-shaped sinkholes mark the spots where the ground has subsided because of solution or where the roofs of caves have collapsed. Sinkholes average ten to thirty feet in depth and may be one hundred feet deep and acres in area; sometimes chains of them merge to make the large depressions called valley sinks.

In a humid climate any such natural basins can become ponds and lakes. Where the water table is high and ground water circulates freely, as in the Florida lake country, nearly every basin may contain water. In localities such as southern Indiana, where the limestone basins have become partly filled with nearly impervious clayey sediments, water can be held above the water table for long periods.

Lake Basins by Wind

The fact that the wind can dig out sizable basins was hard for early geologists to accept, and even today the extent to which this actually happens is uncertain. The power of wind to transport loose sand and soil is attested by the great dust storms of the prairie lands and the sand blizzards

ROCK, TIME, AND LANDFORMS · 266

of deserts. More subtle is the work of the wind as it grinds away the bedrocks by natural sandblasting—a process that is obviously slower than with man's tools, but goes on much longer and more extensively.

Some wind-made basins are blowouts, or hollows fashioned by the wind from loose material among dunes. Others are wind-scoured hollows resulting from abrasion of exposed bedrocks by wind-borne sediments and the subsequent removal of the debris. Lakes occupying wind-made basins are generally temporary; many fall into the class of playa lakes. The water table is well below the bottoms of these temporary lakes; the lake waters consist of runoff from infrequent rains or meltwater from snow on adjacent highlands. If the lake bottom is on sand or on well-jointed, open-textured rocks, the water quickly works down and the lake is soon dry. If the bottom is of compacted clay or other impervious material, most of the water remains at the surface until it is evaporated.

Wind-scoured basins are typically shallow but may cover immense areas. Those in the Basin and Range Province that were filled during the Pleistocene covered thousands and tens of thousands of square miles. Blowouts among dunes are less extensive but often deeper. Thousands of shallow blowouts, holding temporary lakes, dot the Great Plains and the desert and semidesert country from Texas to Montana.

Site of a former lake: Excavations in Meteor Crater, Arizona, show that it contained a lake during humid times toward the close of the Pleistocene.

Water-filled Volcanic Craters and Sinks

A classic among water-filled volcanic craters is Crater Lake in Oregon, the site of ancient Mount Mazama. About 6500 years ago Mazama ceased activity after an explosive period which made the crater a great caldera. This basin, 6 miles wide and 4000 feet deep, now contains 2000 feet of water. Constantly replenished by rain, the lake maintains a level well above the water table of the surrounding terrain.

Frequently lakes form in volcanic explosion pits and subsidence depressions. Well known in the Eifel region west of the Rhine Valley in Germany is the Laacher See (Platten See) near Coblenz, a water-filled explosion pit with a diameter of 1½ miles and a depth of about two hundred feet. France's volcanic Auvergne country has similar lakes, and so does Iceland.

Subsidence depressions are common features in volcanic regions. When the roof of an emptied or partly emptied magma chamber collapses, a depression appears in the ground above it. Probably the best-known water-filled subsidence depression is Lake Balaton in Hungary, which is some 50 miles long, 200 feet deep, and about 300 square miles in area. Other localities with lake-filled subsidence depressions include Hawaii and Iceland.

Lakes in Meteorite Craters

Long considered to be very rare and of little geologic significance are the craters that have been blasted into Earth's crust by large meteorites of the past. Only during the present century have such craters been recognized for what they are. A generation ago just a few were known, such as Meteor (Barringer) Crater near Winslow, Arizona; since then hundreds have been identified.

Sediments forming the floor of Meteor Crater, which is a rimmed basin 3950 feet wide and 570 feet deep, indicate that it was occupied by a lake during postglacial time. Chubb Crater, west of Ungava Bay in Quebec, nearly 2 miles in diameter, is but one of many meteorite craters that are filled with water still.

A great concentration of basins that are possibly of meteoritic origin occur on the coastal plain of North Carolina. Called the Carolina Bays, they have been explained in the past as solution holes.

Some craters formerly regarded as volcanic explosion pits or subsidence depressions are probably meteorite craters instead. Among them is the gigantic water-filled Ries Crater of Bavaria, which is 17 miles wide and 600 feet deep, with a rim rising 100 to 200 feet above adjacent terrain. If this is a meteorite crater, it is the largest known.

A playa in Nevada: A heavy rain can make a broad lake on Harney Playa, but this soon evaporates. The highland in the distance is Steens Mountain.

Lakes Fresh and Salty

For persons who have grown up in fresh-water lake country, salt and alkaline lakes seem positively unnatural. The white incrustations around the shore, the buoyancy of the swimmer (due to the high density of mineral-charged water), and the taste of the water all add to the sense of strangeness. In truth, such lakes are as natural as the sea, and they are salty or alkaline for the same reasons.

Water flowing on and through the crust dissolves many minerals, notably the chlorides, sulfates, and carbonates of sodium, magnesium, potassium, and calcium. These minerals are abundant especially in terrains that were covered by the sea during recent geologic time, and they are picked up by waters that flow into lakes. In humid regions the mineral content of lakes remains low because of frequent replenishment with fresh water and continuous drainage of the mineral-charged water through the lake outlets. But in arid lands replenishment is infrequent, and the ratio of lake water lost through evaporation to water lost through the outlet is much higher. Whereas water leaving a lake through the outlet carries dissolved minerals with it, water that is evaporated leaves the minerals behind; hence salt and alkaline lakes are characteristic of arid terrain. Where chlorides and sulfates predominate, the lake waters are considered salty; where the mineral content is mainly carbonate, they are alkaline.

Because of the current dry trend in the world's climate, the levels of many lakes have recently fallen below their outlets. Such lakes now lose water mostly by evaporation, and virtually all the dissolved minerals that enter them stay there. These bodies of water are becoming, like the sea, saltier and saltier or more and more alkaline, and no end of the process is in sight.

In North America a few salt lakes are found among the salt domes of Texas and Louisiana, and many are encountered in the Basin and Range Province. Over the latter area, with its many diastrophic basins, are exposed salt-bearing sedimentary rocks laid down during past invasions by the sea. During warmer periods of the Pleistocene the depressions were

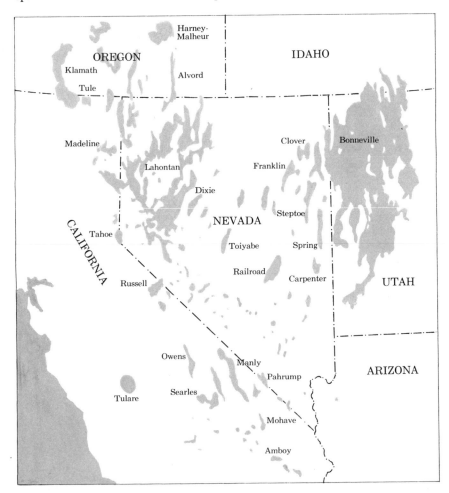

Former lakes of the Basin and Range Province: A region noted for its aridity today sparkled with multitudes of lakes during the Pleistocene.

Shorelines of ancient Lake Bonneville: Wave-cut terraces are discernible on the lower slopes of the Wasatch Mountains east of Great Salt Lake.

filled to overflowing by rainwater and glacial meltwater, and the lakes stayed fresh. The greatest of the lakes was Bonneville, which as already mentioned, at maximum was about 1000 feet deep and covered over 20,000 square miles. This great inland sea drained northward through Red Rock Pass into the Snake River of Idaho. As the climate grew warmer and drier with the close of the Pleistocene, the water supply to the lakes dwindled. Evaporation accelerated, water levels fell below the outlets, and the lakes, with plenty of salt still entering but little or none going out, grew saltier and saltier.

During the ten thousand years since the great glaciers of this region melted, scores of large bodies of water in the Great Basin have completely dried up. Their locations are marked today by extensive salt flats, with terraces where the old shorelines were. Lake Bonneville survives mainly as Great Salt Lake, which covers but 3000 square miles to a maximum depth of only about 12 feet, and it is still shrinking at an alarming rate. The other remnants of Lake Bonneville are Sevier Lake and Lake Utah. The latter has an outlet that keeps it fresh.

Many of the salt and alkaline lakes of yesteryear are now playas. These are the desert depressions in which rainwater accumulates because of poor drainage but soon evaporates under the hot desert sun. A dried-up salt flat with its glittering mantle of minerals is called a salina. A few such basins, including some on the floor of Death Valley in California, are deep enough to reach some ground water and to hold permanent ponds. Death Valley's ponds, being well below sea level, cannot develop outlets and are getting saltier with time.

The Dead Sea, likewise, is salt because of interior drainage. During glacial times its surface was about 100 feet higher than that of the Mediterranean, and its water was fresh. Today the surface is 1400 feet lower, and the water is salt.

The bottom of Lake Baikal in Siberia is 4226 feet below sea level, yet its water is fresh. The reason for this seeming anomaly is that the upper portion of the lake is above sea level, and thus it can have an active outlet—the Angara River.

The mineral contents of salt and alkaline lakes are worth mention. Each 100 pounds of water from Great Salt Lake includes about 17 pounds

A region of dying lakes: In this glaciated area of northeastern Wisconsin, light patches mark sites of glacial lakes that have become filled. Note the lacustrine plain at left.

of dissolved minerals—about 84 per cent sodium chloride (common salt), 6 per cent magnesium chloride, and minor amounts of potassium, magnesium, and calcium sulfates. The mineral content of the Dead Sea, which is even saltier than Great Salt Lake, is 65 per cent magnesium chloride and less than 30 per cent sodium chloride. Chlorides, sulfates, and carbonates especially are all strongly present in the so-called alkaline lakes, among which are Soda Lake in Nevada, Owens Lake and Mono Lake in California, and Abert Lake in Oregon.

The world's saltiest water is found in some of the residue lakes along the edges of the shrinking Caspian Sea. Here the mineral content, chiefly sodium sulfate, approaches 30 per cent of the water by weight. The salinity of ocean water, by contrast, averages only about 3.5 per cent by weight, with sodium chloride making up more than three fourths. The salinity of the Great Lakes and other well-drained bodies is around 0.01 per cent.

Lakes Dying

Barring human interference, most lakes are soon destroyed by natural causes. The overflow from a lake gradually cuts the outlet lower, thus reducing the water-holding capacity of the basin. Plants and animals living in the basin leave their remains on the bottom, dust is blown in by the wind, and other sediments are washed in by running water.

A drying trend in climate makes lakes shrink. Lakes that have accumulated along fault zones, such as the San Andreas Fault in California or Africa's Great Rift, may be emptied when faulting resumes. A lake fed by a stream may dwindle and vanish if the stream is diverted by a landslide, an earthquake, or stream piracy. Nature does not tolerate a lake for long.

From the human viewpoint, lake-filling is an onerous problem. Various unwanted weeds, as well as decorative plants, thrive in lake habitats. Where vegetative cover on adjacent land is cut or bulldozed away, lake filling by the wind proceeds apace. So the early Cape Codders learned after they lumbered away the Cape's original forests and then watched not only their ponds but their harbors fill. The owners of recreational lakes and farm ponds in desert and semiarid localities have had the same problems. It is easier to dig a hole or build a dam than to keep a lake free of choking sediments.

Lakes created by dam building in desert country are especially vulnerable to filling. Frequently the streams that are dammed are overloaded with sediments from the erosion of weak rock. The dam checks the flow and the sediments are dropped. After Hoover Dam was built to make 115-mile-long Lake Mead, the Colorado River began depositing in the lake at a rate that caused consternation among the engineers. Since then the Glen Canyon Dam has been constructed upstream, and Lake Mead presumably has been saved.

Man evolved on Earth just in time to enjoy an extraordinary legacy of natural lakes—the legacy of the Pleistocene. Had we been 10,000 years behind schedule, less than half of the natural lakes of our time would have been here. Today human ingenuity is being applied to the task of converting swamps into farmlands, residential parks, and city pavements—and lakes. Swamps as such appear to be doomed, except where they are saved because of their importance to wildlife. Lakes, on the other hand, appear to be on the increase: natural ones are being saved from nature's own destructive processes, and artificial ones, ranging from the backyard swimming pool to Lake Mead, are multiplying. The growing problem is not how to make lakes but how to find room for them.

XII

Limestone Country

In Yugoslavia's high mountains, fronting the Adriatic Sea, are some of the roughest, most grotesque, yet most beautiful of landscapes. This is the region of the Karst (Serbo-Croatian *kras*, "stone"), a mountainous mass of strongly folded and faulted limestone strata. Rising to 8000 feet and averaging 50 miles in width, it spans the 300 miles and more between Trieste in the northwest and Montenegro in the southeast. The land is grooved, pitted, and cut with steep-walled, ragged gorges, many of them labyrinthine, blind, and dry. Here and there, completely enclosed by mountains, are deep basins which lack surface outlets for the streams and lakes that sometimes occupy them. Beneath all these grotesqueries are some of the world's most spacious caverns and most extensive networks of natural tunnels.

This rugged, intricate land is noted for its barrenness—a barrenness which is the more startling because the rainfall, Europe's heaviest, approaches two hundred inches annually. The anomaly is easily explained. As

The "Parachute": The exact manner of formation of this unique exhibit of flowstone in Lehman Caves National Monument is unknown.

the rains fall, they run quickly into the gullies and pits and are swallowed up and lost underground, not to reappear until they have reached rises, or springs, at lower levels, miles away. Few rivers flow on the surface; most follow underground courses, being visible only where erosion or collapse has removed parts of the tunnel roofs. The mountain slopes exhibit a curious lack of descending river valleys—those familiar features of landscapes elsewhere. Where these valleys do exist they are likely to be dry and blind. This is a world where running water does most of its sculpturing underground.

$CaCO_3$, CO_2, and Water

The Karst is the world's best exhibit of erosion by solution. The land has been shaped by the removal of the calcium carbonate in limestone by water charged with carbon dioxide. Solutional features are developed here so extensively and variously as to become geologic models, and the term "karst topography" is applied to any terrain in which such features are abundant. Although no other region quite matches the Karst, some solutional forms do appear wherever thick, dense limestone is extensively exposed and well drained. They appear also where other soluble minerals, particularly rock salt and gypsum, form bedrocks at or near the surface. But "karst" usually refers to the solution of limestone or of dolomite, its somewhat less soluble cousin.

The chemistry of limestone solution is elementary. This rock, to the extent that it is pure, consists of calcium carbonate ($CaCO_3$). This is poorly soluble in water alone, but when carbon dioxide is added to the water, forming carbonic acid, the acid reacts with the rock to form the compound calcium bicarbonate, which is thirty times more soluble. The water then dissolves the bicarbonate, in the proportion of about 2500 parts of water to 1 part of mineral, and carries it away. In chemical terms the process is represented thus:

$$CaCO_3 + CO_2 + H_2O = Ca(HCO_3)_2$$
$$\text{Calcium carbonate} \quad \text{Carbon dioxide} \quad \text{Water} \quad \text{Calcium bicarbonate}$$

Carbon dioxide is taken by water from the atmosphere and from organic materials, and is already present in the limestone. Hence it is available in abundance, and the erosion of limestone terrains in a humid climate is therefore relatively rapid. In Kentucky the rate of removal of limestone from the land surface is estimated at 1 foot of thickness per 2000 years. As the rock is removed, poorly soluble impurities, such as silica, alumina, iron oxides and carbonates, and sulfur, are left on the land, usually as a reddish or yellowish soil that slows the solution process.

Solution cavities in limestone: These cavities, in rock of the Sandia Mountains, New Mexico, formed by the same processes that create caverns in larger limestone masses.

Dolomite is a limestone in which some of the calcium has been replaced by infiltrating magnesium to form the mineral $MgCa(CO_3)_2$ (this itself is called dolomite). It is less soluble than pure limestone but can yield some karst forms.

Since the prime condition for limestone solution is abundant rainfall, arid regions show little karst development. Such development is favored if strata are elevated; then water will not stagnate in the strata but can pass readily through, carrying the dissolved minerals elsewhere. Drainage is speeded if the strata are well jointed. If the limestone is chalky and poorly jointed, as on England's Downs, water gets trapped in tiny chambers in the rock and solution is retarded.

Intense folding and faulting favor karst development by improving subsurface drainage, as in Yugoslavia and parts of the Greek mainland. Flat-lying strata, however, such as those of the Mammoth Cave region in Kentucky, also show strong solution effects. Rain falling on nearly flat-lying strata moves downward through joints instead of running off via ravines and valleys.

Limestone Sculptures

Exposed limestone surfaces often become honeycombed by rainfall. Runoff on sloping strata can make deep grooves, which follow bands of weaker rock or joints. Grooves may become so deepened that the rock looks like an assemblage of parallel rock fins. The Germans call these karren; the French, lapiés; the British, clints.

Among the most striking forms developed on limestone are the ranks of pinnacles, often gleaming white, seen on deeply incised highlands such as the Bryce Canyon region and the Dolomitic Alps of Italy. These express both high elevation and predominantly vertical jointing. Such jointing favors very rapid downcutting by cascading rainwater and meltwater. Similar effects are seen in the limestone buttresses in the gorge at the Natural Chimneys of Virginia.

Most common among karst features are the sinkholes, which may occur in such numbers as to make the countryside look positively riddled. Central Florida is one well-known example; another is southern Indiana, where the sinkhole total is estimated at 300,000.

The most numerous type of sinkhole, called the doline (from the Yugoslav town of that name), is a deep, funnel-shaped hole formed at the surface where rainwater flows down into a joint or, more commonly, into an intersection of joints. Dolines are usually 10 to 30 feet in diameter, rarely as much as 100 feet. The so-called solution pan, often much wider than the doline, is also shallower. The collapse sink, a depression formed by the collapse of part of the roof of a tunnel or cave, compares to dolines in

Solution work in the Pennines: Karst features in England include these clints, or ridge-and-groove formations, produced by rainwater coursing over limestone surfaces.

A terrain of solution pans and sinkholes in the Everglades of Florida: Such terrains are well developed also in southern Indiana, Kentucky, and Texas.

size but is generally irregular, lacking the doline's smooth, tapered walls. Sinkholes whose bottoms are below the water table, and those partly filled with relatively impervious clay, may contain ponds.

As erosion proceeds on a karst landscape, sinkholes merge to form larger depressions known as valley sinks, or uvalas. These are usually the result of solution and collapse processes combined. They are to be distinguished from the polje, which is similar in appearance but of a different origin. The polje is a closed structural valley produced by downfolding or downfaulting and lacking surface drainage. Pipelike passages through the valley walls, called ponores, pour ground water into the valley or draw it off, according to fluctuations in the water table. Poljes are common in Yugoslavia but not in the United States.

Hide and Seek: Karst Streams

In limestone country, as on the desert, most surface streams are temporary. Only a voluminous river, such as the Ohio where it traverses southern Indiana and northern Kentucky, or the Tarn where it crosses the Causse region in France, can maintain a long course above ground on limestone terrain. On well-drained limestone most stream channels soon lead to swallow holes, which connect with tunnels and caverns below. Swallow holes may drink up rainwater so fast that, as in parts of Yugoslavia, the land is half-desert even where the water table is just below the surface.

On most bedrocks, drainage tends to be two-dimensional; in limestone it is three-dimensional. Channels are often ungraded, because downcutting depends more on solution than on corrasion. The shapes of valleys do not clearly signify stage of development, the influence of rock resistance is obscure. Whereas streams on most terrains become integrated by long-prepared mutual accommodation at the surface, with a nice regard for relative gradients, karst-country streams generally meet underground in the three-dimensional patterns represented by joint systems. Since so much of the stream work occurs below ground, weathering is usually of slight effect.

Areas of erosion in limestone are defined by belts of solution-resistant rock. Limestone strata often grade into shale, and vice versa. Since shale typically is poorly jointed and its particles are small and closely compacted, this rock tends to block circulation. Downward-percolating water spreads out over the shale stratum or follows the dip until it reaches joints or more pervious rocks through which downward movement can be resumed.

Accumulations of insoluble impurities left over from the solution of limestone also tend to block circulation. These fill the sinkholes and clog joints in the bedrock.

A sinkhole in a limestone quarry: This old hole was uncovered near Kankakee, Illinois.

Impervious or insoluble strata sometimes account for the streams called sinking creeks or lost rivers. Such a stream flows over a resistant stratum until it finds a swallow hole. Diving in, it follows a tunnel along a resistant stratum at a lower level. But at a lower elevation this stratum leads to the surface, where the stream reappears as a rise, or large spring. Some sinking creeks disappear and reappear several times within a few miles. They are common in southern Indiana and Kentucky, and one of them is responsible for the great artesian rise in Florida known as Silver Spring. A spring of similar origin is La Fontaine de Vaucluse near Avignon, France, which gives rise to the Sorgue River.

An interesting variant of the sinking creek originates on the highlands of the Istrian Peninsula in Yugoslavia. These highlands consist of a truncated anticline. The stripping of shale strata from the summits has exposed the underlying limestone, but shale survives on the slopes. Rain on the summits forms torrential streams which stay mostly on the surface down to the edge of the shale, but there the waters plunge down into the limestones, follow their underground ways, and emerge miles farther downslope as rises.

Parts of some underground stream courses are visible through what are called karst windows. These surface openings, through which one can look down upon the stream, are especially common in Yugoslavia. They appear to have been formed by the collapse of parts of the roofs of stream-made tunnels.

Karst Valleys and Basins

On ordinary terrain every river-made valley (and this means all valleys except those created by faults) is expected to lead into another valley or out onto a plain that represents base level. In limestone country many a valley ends at a blank wall, often practically vertical, at the foot of which is found a swallow hole. The hole is the point where the river reaches the end of a resistant layer or a junction of joints; from this point the water works downward until another obstacle is encountered. A few such blind valleys are occupied by streams along their full length. Most of them are dry near the blind end; toward the other end the waters have cut far enough down to encounter tunnels.

Dry valleys, which in England are called bournes, may contain streams during wet seasons. During the flood stage the swallow holes may be inadequate to provide quick drainage, so that water becomes temporarily ponded.

Whereas the blind valley ends at a blank wall, the so-called pocket valley begins at a blank wall. A pocket valley forms where ground water emerges near the foot of the slope, dissolving out the rock around and below the point of emergence. These valleys are found typically at the foot of limestone highlands and at the base of the walls of poljes.

The solution valley, or karst valley, results from the solution of strata that are flanked by less soluble formations, such as shale or sandstone. Solution basins, or solution pans, are broad, shallow depressions characteristic of flat limestone lowlands. They are believed to mark sites where the rock is most soluble. As with sinkholes, lakes form in many of these depressions either because the bottom is below the water table, as is true in central Florida, or because it is sealed by clay. A clay cover over limestone may, as in parts of Florida, prevent the development of solution basins and dolines; a cover of sand, which is pervious, will not.

Most lakes in Florida and similar limestone regions are distinctive in that they have not been formed—as most lakes elsewhere have been formed—as a result of faulting, glaciation, landslides, and other such causes. Solution obviously is a great basin-maker. Lakes in limestone, furthermore, can be long-lasting because even while downcutting is lowering the outlets, solution is lowering the bottoms of the basins. But it is also true that those lakes can be suddenly emptied by the opening of a new drainage channel underground.

Cave Country

No karst features have inspired more fancy than tunnels and caverns. With their crazy twistings, fantastic structures of flowstone and dripstone, secret streams and lakes with blind inhabitants that never see the light of day, and tomblike echoes and dank air, these underground worlds have a peculiar hold on the imagination. In this sunless realm prehistoric man found not only shelter from Pleistocene weather but also a place for tribal ceremonies and art work. The attraction of limestone caverns seems equally powerful today for the spelunkers, who, with searchlight and diving lung, crawl and swim through the dark mazes. It is felt even by the tourist who follows a guide along a railinged, cunningly lighted walkway, uttering appropriate exclamations as Victoria Falls, the Cathedral Organ, and the River Styx come into view.

Limestone caverns can develop only where there is good drainage through dense rock. Caves usually start forming at joint intersections, from which they enlarge upward or sideward to become high- or low-ceilinged chambers. Horizontal development tends to be controlled by the bedding planes; vertical development, by the joints.

Whether caves develop above, at, or below the water table was long a point of argument, for some caves are full of water, some partly full, and others empty. It is now generally agreed that solution is most rapid in a zone of circulation just below the water table, where fresh water is contin-

Stalactites and stalagmites in Carlsbad Caverns: These delicate features in the Papoose Room have formed relatively recently at a low level in the caverns.

uously arriving and the mineral-saturated water is being drained away. In cave-forming limestone the water table represents not a continuous surface but rather the elevation to which connecting channels and chambers have been filled.

Streams running over the floors of caves and tunnels may show little tendency to grade them, if solution rather than corrasion is the dominant process. The floors slope up and down and up again erratically. The streams find new routes long before they have graded the old ones.

The existence of so many pockets and blind alleys in caves indicates that only a part of the solution work is done by descending streams. Much is done by the movement of water in all directions—up, down, and sidewise—by hydrostatic pressure. As new water comes down from the surface, the water in all the interconnecting tunnels seeks a consistent level, and its motion never really ceases.

Some caves have several galleries, or stories. These suggest that solution was intensive intermittently, first at one level and then at another successively. Such changes could be caused by uplift or subsidence of the land and a corresponding change in the water table because of changes in downcutting by streams. The water table can be affected also by an increase or decrease in rainfall.

Once a cave has been opened up and the water table has fallen, the cavity can be enlarged by water seeping into it and then flowing elsewhere. Deep vertical grooving on the walls of some caves suggests the importance of this process. Caves enlarge also as blocks of loosened rock fall from the ceiling and are dissolved on the floor. It was in this way that the Big Room in Carlsbad Caverns, New Mexico, acquired its 300-foot-high ceiling.

Potholes in cave floors indicate that streams, too, continue the work of enlargement. Potholes can be made only by waters with a high gradient. Where streams have been ponded by fallen debris, the waterfalls thus produced may greatly speed erosion.

Some caves and tunnels show signs of filling. Streams deposit quantities of clay and other insoluble rock waste that has been carried in suspension. Mineral-charged water dripping or oozing into a cave will, when exposed to the air, redeposit some of the mineral on ceiling, wall, or floor. Some tunnels and caves have been nearly filled with such material.

Caves always represent parts of extensive tunnel networks. Spelunkers like to enter a tunnel from a cave and follow its twists and turns until it opens into another cave, preferably an undiscovered one. Tunnels are generally controlled by both the joint systems and the bedding planes of the rock; they follow a tortuous course that leads up, down, and sidewise as through a three-dimensional maze. Some are properly called siphons, because they lead a flow of water upward from one reservoir and then downward again to another reservoir at a lower level.

Tunnels that suddenly fill with water during a rainstorm become dangerous traps for cave explorers. This hazard is avoided in portions of caves that have been developed for the tourist.

Deposit Features: Flowstone and Dripstone

The distinctive rock of caves is travertine. This is calcium carbonate rock that has been dissolved and then redeposited in a dense, massive, noncrystalline form. Impurities give it hues of red and brown, yellow and orange, and even black. All the forms of travertine that result from dripping are called dripstone; they include stalactites and helictites, stalagmites and columns. Travertine deposited by water flowing over a ground surface is flowstone; it includes terraces and deposits that suggest frozen waterfalls, draperies, organ pipes, and fans. Dripstone and flowstone are often somewhat translucent and thus yield spectacular effects under cave lighting.

Two processes are involved in the formation of cave travertine. First, as mineral-saturated water emerges from cracks and pores of the bedrock and is exposed to the air of a cave, some of the water tends to evaporate, and a corresponding amount of the mineral must therefore be deposited. Second, the emergence of the solution allows some of the contained carbon dioxide to escape, part of the soluble calcium bicarbonate changes back to the insoluble carbonate, and the latter is precipitated out. Thus travertine can form even where the air of a cave is extremely damp and evaporation is minimal.

Stalactites are the iciclelike structures that hang from cave ceilings. They are made by percolating water which, emerging drop by drop from a ceiling, deposits part of its mineral content before it falls to the floor. The deposit takes at first the form of a tiny ring, and this, if the supply of water is steady, gradually lengthens to form a tube. Water descending through the tube deposits new mineral at the tip. Before the stalactite can grow to large size, however, the passage becomes choked and the water flows down the outside, depositing mineral there. The length attained by the stalactite depends upon how far the water can descend before its mineral content is precipitated out.

Where the rate of dripping is sufficient, not all the water evaporates on the stalactite; some falls to the floor and deposits its mineral load there. Thus an inverted cone of travertine called a stalagmite grows up. This may

Helictites: **These unusually fine specimens, formed by moisture gradually exuded by limestone, can be seen at Timpanogos National Monument, Utah.**

build up far enough to join the stalactite and form a column. In caves where the rate of dripping exceeds the rate of evaporation and deposition, the mineral content of the water may be deposited as terraces or other low masses on the floor.

The oddest of dripstone forms is perhaps the helictite, which is formed only in the drier parts of caves. Beginning like a stalactite, the tip soon shows a tendency to twist and curl like toothpaste being squeezed out of the tube. Branchlets may grow out from the sides. With time the helictite may come to look like a confused mass of spaghetti, staghorn coral, or club mosses. This peculiar development is not well understood, but may be due to clogging of the tube opening and to capillary movement of the water sideward and upward after its descent.

Since the underground drainage patterns are constantly changing, the rate of growth of dripstone also changes. The water supply may cease before stalactites or stalagmites can grow large; or it may increase enough to destroy them. Interruptions in the growth of one of these features are likely to be evidenced by the formation of concentric rings in the cross section, like the annual rings of a tree. The rings correspond to surfaces that were weathered, usually to a brownish hue (because of the presence of iron carbonate), between the periods of growth.

The rates of growth for dripstone vary greatly. The largest stalactites

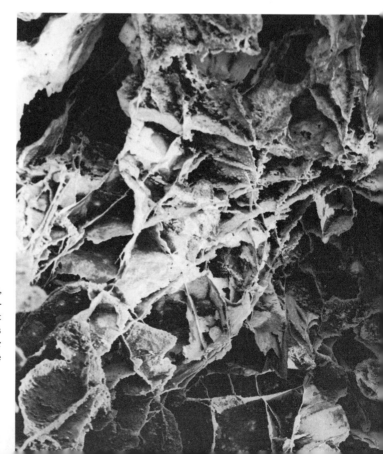

"Boxwork" in Wind Cave, South Dakota: Such structures consist of calcite that was deposited in crevices and remained intact after the more soluble limestone around it dissolved away.

"Scallops" produced by stream-flow over a limestone surface: These specimens of a common feature can be seen at Lehman Caves National Monument.

and stalagmites, such as those in Carlsbad Caverns and Mammoth Cave, are probably tens or hundreds of thousands of years old. A likely average rate of accumulation is a cubic inch per century.

Flowstone in caves is built up by deposits from flowing water. Water escaping from a series of holes in a ceiling or wall may produce vertical columns resembling organ pipes. Water flowing out of a long crack can create a drapery, a waterfall, or a cascade of stone. At the edge of a cave pond, deposition of flowstone may build up various kinds of terraces, including the popular scalloped forms called lily pads. Some terraces become retaining walls for ponds. Where water moves sheetlike over a floor, a travertine pavement is made. Interruptions in flowstone growth may create striking banded effects in the stone.

Crystalline forms of calcite are built up by the very slow emergence of solutions from pores in the rock. Entire walls and ceilings may become covered with tiny crystals which, with artificial lighting, sparkle like the star fields of the Milky Way. These crystals are formed somewhat like the "frost flowers" that rise from moist soil during a freezing night. Elsewhere in a cave one may find large expanses covered with travertine in spongelike masses or clusters of long, needlelike crystals resembling frost on a window or thin, new ice on a pond.

Gypsum rosettes: Such calcite growths, found on cave ceilings, are common in Mammoth Cave, where this photograph was taken. The mineral is deposited by seepage.

Among the most exquisite cave deposits are the usually pure white gypsum rosettes. These are not travertine but hydrated calcium sulfate ($CaSO_4 + H_2O$), a soluble mineral formed from calcium carbonate and sulfur-containing compounds. Emerging from pores of the rock, the gypsum-bearing water forms needles, buds, and blisters, some of which grow and open up to form the rosettes. These require growth conditions different from those of dripstone; they are found only in dry areas, and never in company with stalactites and stalagmites. They may grow to a diameter of several inches.

Although in geologic terms the solution of limestone is rapid, by human standards it is very slow. The rate of over-all reduction of the Mammoth Cave region of Kentucky was mentioned earlier as about 1 foot per 2000 years. By this standard a cave zone 250 feet in depth might be expected to develop over a period of 500,000 years. Study of the Lost River cavern in southern Indiana indicates that it has formed since the early Pleistocene, perhaps a million years ago. The upper levels of Carlsbad Caverns in New Mexico contain washed-in pebbles from sedimentary rock strata dating from the Pliocene Epoch, 1,500,000 or more years ago; therefore, these levels may be that old.

Some caverns, such as those of the Ingleborough district of England, have been dated by means of glacial debris washed into them. The debris came from surface deposits which have been dated by stratigraphic and carbon-14 techniques.

Probably the large cavern systems of the world have taken from a half million to several million years to form, and a few may exceed several million years in age. The highest levels of the caverns are the oldest, and the deepest levels the youngest.

The minor ornamental features in caves are younger than the caves themselves. Dripstone and flowstone build up only as the water table falls and the cavern's surfaces become exposed to the air. Then these features can grow only as long as the necessary moisture conditions are present.

Large stalactites, stalagmites, and flowstone in the upper levels of a cave that has long been dry are very ancient—almost as old as the cave. But crystalline forms of travertine and gypsum on ceilings and walls at the same levels may be of very recent growth. Since weathering in caves is much slower than the weathering of rocks outdoors, both dripstone and flowstone features may persist unchanged for millennia.

The Karst Cycle

A karst terrain may undergo a distinctive kind of evolution. This can begin with the uplift of an extensive limestone formation that has been near base level, with consequent rejuvenation of its surface drainage. Or the cycle may start when the stripping of insoluble rocks such as sandstone or shale from a highland exposes underlying limestone strata. In either instance drainage on and within the limestone is accelerated, and the typical karst surface features soon appear—lapiés and sinkholes, solution valleys and solution pans, and the rest.

If the limestone is covered with insoluble rock through which streams have cut far enough to reach the limestone, karst valleys will develop along the stream courses. The valley walls are formed by the resistant strata, and the valley bottom is in the limestone. If the valley bottom represents a local or temporary base level, surface drainage will predominate, perhaps with meandering streams and marshy or poorly drained areas. If there is good drainage from the valley to still lower levels, karst features will evolve.

In the youthful stage of the cycle, systems of tunnels and caves are opened. Surface streams evolve into sinking creeks and lost rivers, and most of the surface drainage features disappear. Here and there the collapse of a section of a tunnel's roof makes a karst window. Sinkholes coalesce to form valley sinks, and valley walls are carved to form steep-walled dry ravines and gullies, with perhaps occasional pocket valleys along the base. These developments bring the landscape to maturity, the state of maximum ruggedness and maximum slopes. Maturity lasts until the region is reduced nearly to base level or until the percolating waters reach strata that are less soluble or poorly jointed.

Then old age begins. Subsurface drainage slows; surface drainage is gradually restored. The pits and gorges are converted into a landscape of open valleys, and the ridges become small, isolated hills. (These conical, solution-riddled monadnocks, best developed in humid tropical lands, are called haystack hills in the United States, pepinos in Puerto Rico and Jamaica, hums in Yugoslavia, mogotes in Cuba, and buttes temoines in the Causse region of France.) Also in this old-age phase, natural bridges may take shape as surface erosion unroofs tunnels used by underground meander cut-offs, or tunnels through which underground streams were drained by pirate neighbors. And so, with the restoration of surface drainage, karst forms disappear, to be replaced by a standard topography of insoluble rocks.

The Karst Regions

Limestone is a common rock, occurring in formations ranging from local lenses only a few inches thick to strata covering thousands of square miles to depths of many hundreds of feet. Every continent has extensive limestone terrains. A tourist in the United States may get the impression that every state claims the world's finest crystal caverns. Actually, areas of cavern development are more common than areas in which other karst features have developed.

The United States offers good karst scenery in central Florida, the

The Causse near Meyrueis, southern France: The bleak land suffers not from lack of rain but rather from the rapid escape of rainwater through joints of the limestone into underground channels. So with the karst landscapes of Yugoslavia and Greece.

Great Valley region of Virginia and Tennessee, and above all the region of southern Indiana, west-central Kentucky, and north-central Tennessee. The Mammoth Cave locality of Kentucky exhibits virtually all the karst forms.

Mammoth Cave is in a region which, about 260 to 270 million years ago, during the Mississippian Period of the Paleozoic Era, was a broad seaway extending northward from the Gulf of Mexico. Calcium carbonate, either precipitated from sea water or derived from the remains of shellfish, corals, and other lime-containing organisms, accumulated on the sea bottom, eventually forming strata about 1200 feet thick. For millions of years these deposits were covered over by shale and sandstone wastes from the Appalachian Mountains to the east. Then the whole area was uplifted to form the Cincinnati Arch. Accelerated erosion stripped off much of the shale and sandstone, exposing the limestone. It is in this limestone that Mammoth Cave has formed.

This cavern is in the western flank of the arch, where the stripping is still incomplete. North of the cave region the limestone is protected by the Pottsville sandstone. Beginning at the southern edge of this formation and at a lower level is the Mammoth Cave Plateau, from which the Pottsville formation has been removed completely. Exposed here are the Cypress shale and sandstone strata, through which streams have penetrated far enough to reach the underlying limestone. Water percolating through the limestone drains northward, following the dip of the strata—about thirty feet per mile—into the valleys cut by the Nolin and Green rivers. It is between the surface of the Mammoth Cave Plateau and the Green River that the caves have developed. The Green River valley is the outlet that ensures the circulation vital for cave development.

The southern edge of the Mammoth Cave Plateau forms the well-named Dripping Springs Escarpment, which rises above a lowland known as the Pennyroyal Plateau. Whereas the surface of the Mammoth Cave Plateau is still partly protected by the Cypress shales and sandstones, and shows few karst features, the surface of the Pennyroyal Plateau to the east is composed completely of limestone, and is so pitted with sinkholes that it has been called the Land of Ten Thousand Sinks. But cave development here is poor. On both plateaus surface streams are rare, although some patches of the Cypress shale contain ponds.

Mammoth Cave consists of a three-dimensional maze of caves and tunnels developed within an area of about 80 square miles with depths far exceeding the 360 feet that is the limit for tourist excursions. Scores of miles of these passageways have been explored. Five different cave levels are distinguished, each representing a zone in which solution was more rapid than in the strata immediately above and below it. Corkscrew passageways, some of them partly blocked by rockfalls, connect the adjacent levels. At joint intersections, great domed chambers have been developed,

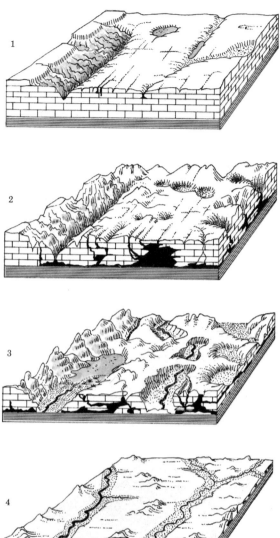

The karst cycle: Surface drainage is erratic and poorly developed throughout the cycle, becoming normal only with the reduction of the landscape to the level of rocks other than limestone.

and broad, low-ceilinged rooms have been hollowed out along the bedding planes. The upper levels of the cave are relatively dry; the lowest levels, wet. At the bottom level, 360 feet down, flows Echo River, which receives most of the downward-moving waters of the cave before it joins the Green River.

The cave has chambers of awesome size. Mammoth Dome, 192 feet high, connects with Mammoth Dome Sinkhole at the surface. Presumably the dome was formed by the solutional action of waters funneled down through the sinkhole. The Silo Pit, another solutional chamber, is 160 feet deep; it was formed at an intersection of joints. The Great Rotunda Chamber, 139 feet wide and 40 feet high, originated by solution and the falling of thin limestone slabs from the ceiling.

An unusually impressive rockfall is represented by Giant's Coffin, a block 50 feet long, 20 feet wide, and 16 feet thick. The sight of this block gives the tourist uneasy thoughts until he is told that such rockfalls are extremely rare, and no visitor in the cave has ever been injured by one.

Petrified waterfalls are represented by Frozen Niagara, a mass of flowstone 75 feet high and 45 feet wide, deposited by water dripping and flowing over a heap of fallen blocks. In the Drapery Room is an array of curtains formed by the joining of stalactites. The Great Wall of China is a travertine dam that ponds a stream, and Crystal Lake demonstrates how a stream can be backed up by the growth of a stalagmite. A particularly beautiful display of curtains, stained yellow by iron oxide, with a multitude of young, dripping stalactites, appears in Onyx Chamber.

The ceiling of Star Chamber is a glittering firmament of white gypsum crystals that have grown through a surface stained dark by manganese and iron oxides. Along Silliman's Avenue are exhibits of gypsum rosettes, and of "grapes" formed by suspended water drops on the ceilings. The Snowball Dining Room, where tourists have lunch, has a ceiling plastered with "snowballs"—actually, gypsum blisters that have not opened to form rosettes.

A favorite feature is the River Styx, upon which one takes a short boat ride. This is one of the smaller streams. When the water supply is greater than usual, this river pours the excess into Echo River.

Many caves elsewhere rival Mammoth in scenic interest. Some have been opened in limestone hills from Virginia to Pennsylvania along the Great Valley, between the Blue Ridge Province on the east and the Appalachian Ridge and Valley Province on the west. Isolated caves are encoun-

Pepino hills in Puerto Rico: These remnants of a much reduced limestone terrain have the typical peaked aspect. In the foreground is the coastal plain.

tered in New York state and New England. Besides the caves in the well-developed karst regions of southern Indiana and Tennessee, there are spectacular specimens in Missouri, Arizona, and California. Oregon has its marble caves, and South Dakota its Wind Cave. In Colorado an unusual feature is the very dry Cave of the Winds, near Manitou, developed within a limestone cap on a hill of crystalline rock.

Among the finest caves are Carlsbad Caverns in southeastern New Mexico. They are covered by flat-lying, nonsoluble sedimentary rocks that give no hint of the underlying maze of chambers and tunnels. These were formed by solution in the great uplifted fault block of the Guadalupe Mountains, mainly during the wet periods of the Pleistocene. Today the Carlsbad complex of open caverns and tunnels reaches to depths of over 700 feet, and there the water is still working. The Big Room, which is nearly 4000 feet long and 600 feet wide, with a ceiling almost 300 feet high, ranks as the largest solutional chamber in the world.

The karst region of northern Yucatán, jutting out into the Gulf of Mexico, is one of the most fully developed in the Americas. Here a gently dipping limestone plain is riddled with sinkholes, known locally as *cenotes*, and beneath the plain are extensive cavities. The water table is at a depth of 60 to 70 feet inland but rises nearer the ground surface as one ap-

Chalk formations in Providence Canyons near Lumpkin, Georgia: Water trickling down old Indian trails made gullies which in time became canyons. The largest of these is now one-half mile long, 150 feet deep, and 300 feet wide.

proaches the coast. Water follows the dip of the bedrock, running seaward between the strata. Finally it emerges offshore as artesian springs rising from the bottom of the Gulf.

The central areas of Florida display many sinkholes like those of Yucatán. Silver Spring, east of Ocala, is one of the most famous of artesian wells, its flow averaging about 500 million gallons per day. Florida is noted also for its sinkhole ponds and solution pans.

In the Caribbean, the coral-capped volcanic islands of the West Indies, with their humid climates, rank high among karst exhibits. In northern Puerto Rico, plains representing late phases of the karst cycle are unusually rich in haystack hills. The "cockpit" country of Jamaica is noted for its sinkholes. In England, Yorkshire has striking karst features in the Ingleborough region, including dry valleys and the famous Gaping Ghyll, a swallow hole that leads down 365 feet into a chamber 480 feet long and 110 feet high.

Notable karst areas of Europe include also those of Greece and, especially, the Causses region of south-central France. Unlike karst topography in the United States, that of the Causses has developed on and in rocks that are folded and faulted, as in the classic Karst country of Yugoslavia. Deep river-cut gorges divide the land into blocks, the surfaces of which are pitted with sinkholes, some of them as much as 700 feet deep. Underground drainage is well developed; dry valleys are the rule, and the sight of a surface stream is rare. As is usual on karst terrain, agriculture here is difficult. It was in the caves of the Causses, such as those of Lascaux, that prehistoric man left his priceless paintings.

A shore of submergence: The relative rise of sea level in recent time has subjected many high-standing, rocky terrains to marine erosion. At Hartland Point, Devon, England, waves of the Atlantic beat against cliffs 350 feet high. Note the razorback shapes of the offshore islands, due to erosion of stratified rock.

XIII

Coasts and Shores

Seashores are rich in diversity and rapid to change. Rugged cliffs fronting deep water along America's Pacific coastlines contrast with the low, sandy Middle Atlantic shores, from which shallows stretch miles offshore. There are coastlines long and straight, like those of California, and others indented deeply by estuaries, as on the coast of Maine. Many shorelines, like Long Island's, are scalloped with shallow coves and bays, or fringed with spits, bars, and islands. There are sandy beaches, gravel beaches, cobble beaches, black beaches. Some shores, like those of Cape Cod's Atlantic side, are visibly retreating, losing a few inches or feet—sometimes even a few yards—in a heavy storm, often to the vexation of cottage owners and beach engineers. Other shores, like Cape Cod's extremities, are steadily building seaward and, it is to be hoped, making someone rich. But all these zones where land meets sea are alike in being shaped by moving water—the waves that beat upon them, the currents that drag the sands along them.

The continents were described in earlier pages as blocks of light rock "floating" in the slightly heavier basalt of Earth's crust. Their margins are the approximate edges of natural basins formed by inequalities of density in the crust. Coasts are zones where erosional and diastrophic changes are especially noticeable. Shore scenery usually reflects current phases of uplift or subsidence, and in fact but slight changes in elevation can work abrupt scenic changes.

Whereas erosion inland consists mainly of downwasting, shore erosion is primarily a horizontal cutting—a kind of sideward sawing into the land. Waves and currents work by striking and wedging, bulldozing and heaving. Like streams they also use solution, hydration, and corrasion, with the help of weathering and gravity, and they are continuously transporting and depositing erosion debris.

The over-all character of many shores has been determined by events of land rather than marine origin. Coastal contours may result from the invasion of river valleys by the sea, as in Maine, or from underground solution, as along Yucatán's limestone shore. Other areas that front the sea, as in Hawaii, show features due to volcanism. Shores on the Gulf of Mexico have been shaped largely by the deposit of river sediments, and the coast

A shore in the Virgin Islands: Coral growths around these volcanic islands have supplied the materials for white beaches. Beaches tend to be narrow because of the steepness of the offshore slopes.

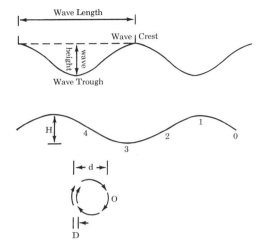

Wave mechanics: The land-shaping capability of waves depends upon the forward movement of the water particles and the breaking of waves as they enter shallows. D is the progress made by a water particle O while the wave form is traversing the distance d. H is the wave height.

of eastern Massachusetts by the accumulation of glacial debris. Wind-blown salts and sands make up shores around bodies of water such as Great Salt Lake and the eastern Mediterranean Sea. Finally, there is coastal relief that expresses uplift or subsidence primarily, as in southern California.

Other shores show the influence of marine rather than terrestrial agents. The coasts of Washington and Oregon, for example, are distinctive in having cliffy shores of homogeneous rock that have been straightened by long wave and current action. The span between New Jersey and Georgia expresses with different relief the work of the same agents. These are typically low coasts with broad beaches, mud flats and salt marshes, and spits and offshore bars. Finally there are coasts sculptured in organic deposits, such as those of mangrove in western Florida and coral or coralline algae in the islands of the West Indies and the Southwest Pacific.

The Anatomy of Waves

The most spectacular of all ocean waves is the tsunami, sometimes inaccurately called a tidal wave, which results from submarine faulting or a volcanic eruption. It was such a wave, started in Sunda Strait by an explosion of Krakatoa Volcano in August, 1883, that swept up the shores of Java and Sumatra as a 100-foot-deep flood, overwhelming villages and drowning some 35,000 persons. Waves from the Krakatoa explosion were still several feet high after traveling thousands of miles across the Indian Ocean. Tsunamis with heights up to 60 feet, set off by a sea-bottom fault in the Atlantic on November 1, 1755, destroyed much of Lisbon, Portugal. Less tragic but locally more spectacular was the wave built up in Lituya Bay, Alaska, on July 9, 1958, when some

40 million cubic yards of soil and rock, shaken loose by an earthquake from the 4000-foot side of the fiord, plunged down into the water. The water rose 1700 feet against the opposite side and swept outward from the bay as a 50-foot wave traveling about 100 miles per hour. It heaved a fishing boat entirely over the spit at the mouth of the fiord and sank it.. Such catastrophic waves have left marks on many coasts.

Much smaller, but more destructive in the long run, are wind-made ocean waves. These originate mostly in deep water far from shore. Waves form as the wind pushes down on the water, and are built larger as the wind pushes them from behind and sucks at their leeward sides. Strong winds make white caps by blowing off the tops of the waves. Apparently waves are increased in size by resonance; that is, the wind and water interact rhythmically, like the child on a swing and the friend who is pushing him higher and higher. Wind-generated waves, often formed in a confusion of gusts, are complex. They are rarely very large, but being multitudinous and ever moving, they are effective shore-shapers.

Height and period are important wave properties. Height is the distance between crest and trough, and period is the time span between the passage of two successive crests by a given point. Height and period depend upon the wind's velocity, fetch (distance over which it blows), and duration. Far offshore, waves often reach heights of sixty feet, and during storms in the North Atlantic veteran observers have judged some waves to be seventy-five feet high. These giants are believed to require a fetch of five hundred to six hundred miles and two days or more of blow.

The wavelength, or distance between successive crests, also depends upon fetch. It ranges from a maximum of twenty feet on small lakes, where the fetch may be only a few hundred yards, to two thousand feet on the oceans of the southern hemisphere. There the winds blow unchecked over thousands of miles of watery waste.

Each water particle in a deep-water wave has a nearly circular vertical motion. As water at the surface is pushed forward by the wind, water from below rises to replace it. Thus each particle moves forward, downward, backward, up again, and forward once more. With each circuit there is a slight forward gain, and so the waves as a group show a steady leeward movement, or drift. As wind-blown water moves out of the region, other waters move in; thus wind can create currents. Wave drift is much slower than the progress of the wave form, as any fisherman will notice when waiting for waves to bring his loose boat back to shore. But the water in waves may nevertheless travel thousands of miles.

The ocean waves that travel farthest become what is known as swell. These are long-period waves that move faster than the others and are the first to reach a shore distant from the area of wave generation. Because Earth's rotation deflects the trade winds eastward in the northern hemisphere and westward in the southern hemisphere, long-period swell is noticed more on the western coasts in the northern hemisphere and on the eastern coasts in the southern hemisphere.

On all shores incoming swell is continuous, varying somewhat in magnitude according to local wind conditions. Besides swell there are also wind-backed waves which have been generated near the coast; these vary greatly according to the strength, direction, and persistence of the wind. Both swell and locally generated waves enter into the making of breakers, which are the strongest of the attacking agents along shores.

The destructiveness of breakers is due to their nature as waves of forward motion; that is, waves in which the circular motion of the water particles has been converted into translational motion. As long as waves are in deep water, they are easily reflected from an obstacle, with little loss of energy, by simple reversal of the circular motion. But where a wave comes in over shallows that slope gradually up to the shoreline, bottom drag interferes with the circular movement, and reversal cannot occur. The velocity and wave-length of the wave decrease, its height increases, and the water in the top of the wave finally plunges over the slowed water beneath it. Thus a breaker is born.

Waves of the long, low type break when they reach a depth about equal to their height. Short, steep waves break along a line where the

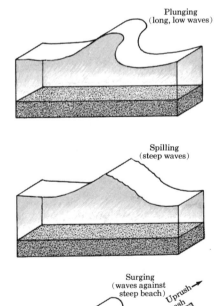

The origin of breakers: The formation of these waves results from bottom drag on the advancing wave form.

depth is a little more than twice their height. Because of bottom drag, incoming breakers become relatively high where they are following a ridge along the bottom and relatively low where they are following a trough.

Plunging breakers—those that have the greatest impact on cliffs and sea walls—tend to be produced by long swell coming in over shallows against opposing winds. The spilling breaker, which crashes gradually, is typically produced by steep, wind-backed waves of short period. The surging breaker, which floods rather than breaks, is common where the beach slope is steep.

Waves against the Land

Plunging breakers from the largest storms strike cliffs and sea walls with a force as strong as seven tons per square foot. Thus sea walls can be breached and lighthouses swept away. Ordinary waves, on the other hand, make up for their lesser strength by greater persistence. They plunge against sea cliffs day after day, century after century. As water rams into the joints, it compresses air so as to wedge rock masses apart; and as it withdraws, the trapped air expands explosively. Meanwhile, constant washing decomposes the rock by hydration and solution, and above the water line weathering agents also work to disintegrate the cliff.

On a steep shore the destruction is greatest in a zone along the water

Breakers on the shore of Montague Island, Alaska: Spilling, breaking, and surging waves are discernible along this mile-long portion of the shoreline.

A sea cave and a notch: Waves and currents at Acadia National Park, Maine, have cut a cave (*left*) in a zone of weak rock and the notch near the high-tide line.

line, and so a wave-cut notch typically develops at that level. The undercutting produces overhangs that periodically break off and fall. As the cliff recedes, undercutting may keep it steep.

The rock-destroying power of waves is far less below the water line than above it. A wave-cut rock platform, or bench, therefore, usually develops underwater as a cliff retreats. Rock waste on the bench is picked up by the waves and hurled at the cliff's base. With time, as the bench widens, erosion debris accumulates on it to form a narrow beach. Growing outward, the beach increasingly breaks the force of the incoming waves, and undercutting of the cliff ceases. Accordingly the cliff, now slower in retreat, acquires a gentler slope.

Just shoreward of the breaker zone the bottom takes a heavy pounding. The uprush, or swash, of water carries with it a highly corrasive load of sand, gravel, and cobbles. During moderate weather the waves disturb only the upper layer of the debris that covers the bench, but in storms the entire mass may be moved back and forth, grinding the bench lower.

Waves are especially determined in their attack on a headland. Those moving parallel or nearly parallel to the main shore hit the side of the headland hard. Also, waves passing near the sides of the headland are slowed by friction with the bottom and thus tend to swing against the shore. This familiar refraction phenomenon causes headlands to be eroded faster than the rest of the shore.

The cliffy shore of Acadia National Park: Here a beach is in the making, but it is a rough one. Sand from inland is lacking, and the waves have had insufficient time to break up rock fragments from the cliffs.

Waves work and rework the debris of erosion. Gradually it is reduced by impact, rubbing, hydration, and solution to form the materials of beaches. Much becomes trapped in coves and bays. Some is moved back and forth between the shallows and deep water.

All material within reach of waves tends to migrate along shore in the process called beach drift. It is carried up the beach at an angle by oblique swash and then swept down again, perpendicularly, to an advanced position. Beach drift is generally in the direction of the dominant winds, except where refraction may turn it somewhat to windward. The long-time tendency of the sediments is to migrate downwind and accumulate against other shores far away.

Currents along Shore

Working with waves along a shore are the various kinds of currents. In order of increasing importance these are the ocean currents, which form out at sea; the tidal currents, resulting from the ebb and flow of the tide; the longshore, or littoral, currents, produced by wave action; and the rip currents, consisting of water that is returning seaward after having been heaved against the shore.

All oceans are laced with wind currents—those generated by prevailing winds. Also formed at sea (and near shore as well) are the density currents, which are due to temperature and salinity differentials. Cold water, being denser and heavier than warm water, tends to sink beneath it. A mass of water containing a larger amount of minerals in solution than an adjacent mass will act similarly. Thus the waters of the ocean at depth are constantly moving. Although ocean currents, both wind-generated and density types, do little erosional work on shores, they influence shore cur-

Current patterns off the Georgia coast: The light, featherlike areas are sand deposits made by longshore and rip currents. The ribbing (*lower left*) is a boat's wake. Fainter lines to the left of the wake are waves.

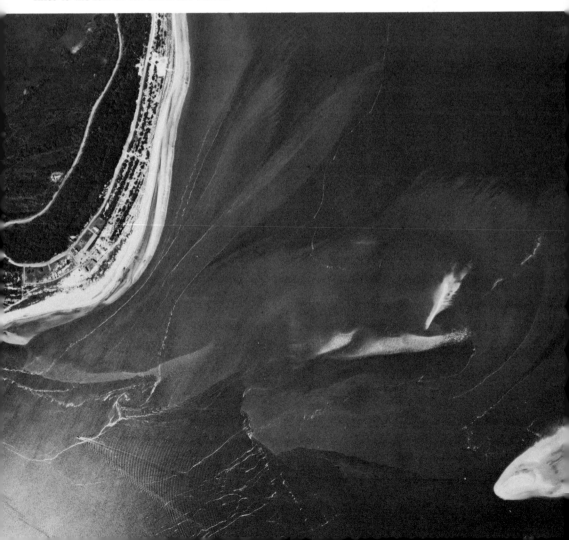

rents and also undoubtedly transport substantial volumes of sediments.

Tidal currents work strongly along embayed seashores. Tidal water sliding along the flanks of headlands and along the fringes of bays and coves picks up and redistributes small-caliber sediments. When such currents enter inlets through narrow channels, constriction speeds them and increases their erosive ability. If a narrow channel has just the right funnel shape, it may form fast-moving tidal water into the high, translational wave called a tidal bore. In the Bay of Fundy, where spring tides have a range up to sixty feet, the bore reaches a height of six feet and the tidal currents attain speeds of fifteen knots. In Seymour Narrows between Vancouver Island and the mainland of British Columbia, speeds of ten knots have been measured.

Waves approaching a shore obliquely not only cause beach drift but may create powerful longshore (littoral) currents, which move more or less parallel to the shoreline. Such currents, sometimes reinforced by tidal waters or ocean currents, can be more erosive than waves and may become the principal architects of a shore.

Rip currents—sometimes loosely called rip tides, or sea puss—consist of water that has been thrust against the shore as translational waves, converted into longshore currents, and then turned seaward again. Characteristically, rip currents are narrow flows extending all the way from the surface to the bottom, and reaching hundreds of feet—rarely thousands —to the rip head, or seaward end, where they die away. Rip currents develop in series, the spacing depending upon the contours of the bottom, the slope of the shore, and the height and period of the waves. Where a rip current has formed, breakers are usually weak, the water is agitated into small waves, and sand is being churned up from the bottom.

With their speed of a mile per hour or more, rip currents are efficient carriers of sand and gravel and can deplete beaches, especially in winter when waves are high. Sometimes they catch swimmers and toy boats, carrying them rapidly out toward deep water. The experienced swimmer or boatman caught in a rip current does not try to resist it, but makes his way sideward to get out of it. "Undertow," popularly believed to be a current of water returning seaward beneath the incoming waves, and blamed for the loss of many swimmers, exists rarely if at all. Bottom currents are much more likely to be moving landward than seaward. Backwash, however, can upset a swimmer and sweep him into a rip current.

Erosion by waves and currents is concentrated mostly within a zone only a few feet in depth. It is slight at depths greater than thirty to forty feet, though swell has been known to move lobster pots at depths exceeding one hundred feet. Notwithstanding, the zone of activity is deep enough to keep the edges of the land in retreat unless counterevents occur—uplift, for example, or the addition of land mass to the coast by delta-building, lava flows, or deposits of glacial debris.

COASTS AND SHORES · 307

Shores Through the Millennia

Although, for reasons soon to appear, few if any shores ever undergo a complete erosion cycle, the work of waves and currents does imply the possibility of such a cycle. Marine erosion actually is changing every shore

The chalk coast of Normandy, France: Waves and currents have cut out the arches.

according to the logic of certain mechanical and chemical processes. Every shore regardless of its origin can be regarded as being in some stage of a cycle—early, middle, or late—even if the cycle will never be completed.

To demonstrate some possible patterns of evolution, we can begin with a high cliffy terrain which, by uplift or subsidence, has recently become subject to attack by waves and currents. Either the coast of southern California, which has been uplifted, or that of Maine, which has been depressed, would serve as a model. Waves have been attacking the cliffs, cutting the notch at the water line and developing the wave-cut bench beneath it. Plunging waters are searching out the zones of weaker rock, opening joints to form inlets and perhaps hollowing out tunnels and caves. As tunnel and cave ceilings collapse, openings appear in the ground above, like the collapse sinkholes of limestone country. During storms, fountains of water burst through some of these openings—which are, appropriately, called blowholes.

Gradually the cliffs retreat. The rock masses that are relatively resistant or that were originally partitions between joints become short promontories, and as these are cut through they form sea arches. When the tops of the arches collapse, the sides remain as the offshore islands, called sea stacks.

With further erosion the shore begins to show the small, crescent-shaped indentations called bights, and the deeper ones known as coves. Inlets are formed that connect with the sea through narrow channels. Also taking shape are the larger indentations called bays, with their wider seaward openings.

Such irregularities are favored where the shore rock varies in resistance and structure; but where the variations are minimal, the sea works to

An aging Pacific shore: In Olympic National Park, the sea has been at work long enough to dissect a high-standing terrain and make a wide beach.

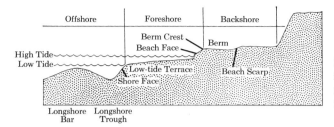

Profile of a beach: All these features are common, but many beaches do not exhibit them all.

straighten the shore. Headlands are attacked with especial vigor as waves are refracted against their sides, and coastal indentations tend to be widened, rather than deepened, by similar action against their sides. Meanwhile drifting rock debris becomes stalled around the inner margins of bays and coves, and is supplemented by sediments brought from inland sources by wind, streams, and mass wasting. Longshore currents and beach drift distribute the rock wastes along the shore and, in doing so, reinforce the straightening process.

Rock waste accumulating on the wave-cut bench below the cliffs gradually rises above water to form beaches. These can form fully only during a long period of stability. They may appear along an open shore, especially where it is nearly perpendicular to the prevailing winds, but usually develop first in shore indentations as pocket, crescent, and bayhead beaches. Here sediments carried along the shore by wave and current action become trapped.

The materials of which beaches are made—silt, sand, gravel, cobbles, and shingle—are derived from sources as various as lava, coral, glacial drift, delta deposits, and granite. On relatively young beaches the sediments may be mostly of large caliber, because rock fragments are falling from the cliffs faster than they can be ground up, and because fine sediments are moved more easily seaward. If the rock is generally homogeneous, as granite would be, wave work tends to round the fragments; but if the rock is sheety, like shale or schist, the fragments remain flattish and hence are called shingle. Beaches loaded with coarse sediments are leveled toward the horizontal but slowly, because the swash drains through this material readily and there is little backwash.

Full beach development demands an adequate supply of sediments. This means that sediments produced by erosion must not be lost too fast by downshore drift or by gravitation toward the deep sea bottom. Seashores of highly resistant rock and steep seaward slopes, as in southern California, may show little beach development, because the sediments are sparse and soon lost. Beach development is extensive from Long Island southward because of the gentle bottom slope and the bountiful supply of glacial drift and of sediments from the erosion of weak rocks inland.

All beaches show some if not all the features included in the accompanying diagram. A beach is generally defined as the zone of loose erosion debris between the upper limit of wave action and the lower limit of the tides. At the landward limit of the beach is a cliff or, if this has been worn low, a gentle slope beyond the reach of all but the highest storm waves. This slope is technically known as the coastline. In front of it is a gentler slope leading to the beach scarp. This is a low, steep incline extending to the berm, a nearly horizontal platform reached by storm waves in times of high water. The berm and the slope just behind it together form the backshore.

In front of the berm is the foreshore. This is bounded on the landward side by the beach face, a steep slope in front of the berm, marking the normal limit of swash at high tide. Just below the beach face is the low-tide terrace, a gentle slope exposed at low tide. At its seaward limit a steep slope called the shore face marks the landward boundary of the offshore zone.

On the foreshore one finds swash marks: thin, low ridges of sand, seaweed, and other light detritus left by uprush. Here also appear sand ripples, both on the exposed beach and underwater. Such ripples are formed by waves and currents.

The small drainage channels running downslope on a beach, made by backwash and by water draining out after high tide, are called rill marks. Ridges between the larger channels may be shaped at their lower ends by the waves to form beach cusps—short, low mounds with points on their seaward ends. At low tide, especially on the lower beach and on exposed longshore bars, rip channels may be seen: these are trenches dug out by rip currents.

At the foot of the shore face is the shore trough, formed by the digging and gouging action of breakers. Beyond this is the longshore bar, a parapet of sand and gravel thrown up by the same action. The waves break over this bar, which is often visible at low tide. The position of the bar may change frequently. The larger the waves, the farther out the bar forms.

Such a profile is by no means universal. Beaches in an early phase, particularly gravel beaches, usually have a continuous slope from the backshore (the sea cliff or the coastline slope) all the way out to the shoreface, or low-tide line. Later a berm is likely to develop, and then the other subdivisions of the complete beach.

On some beaches storm waves have cast up gravel and cobbles behind the berm to form one or more parapet-like structures called beach ridges. These are to be distinguished from dune ridges, which may form behind the beach and are of finer material. A beach ridge may last for years before the waves of a great storm break it up.

The marine-cut terrace, sloping seaward beneath the beach, represents the deepest penetration by wave and current action. Its inner portion is the wave-cut bench, and its outer part the abrasion platform, across which sediments are continually being dragged by waves and currents. In summer, the sediments tend to build up on the terrace, but they are dug out severely in winter.

Seaward from the abrasion platform may be a marine-built terrace, covered with sediments that are disturbed but rarely. Off some coasts the abrasion platform and the marine-built terrace together constitute the continental shelf—a surface that may be wide or narrow, flat or hilly, sloping or horizontal. From this shelf the continental slope drops to the deep ocean floor.

By the time the original sea cliffs have been worn low and beaches with longshore bars have developed, various other depositional shore features will be taking shape, provided the sand supply is adequate. One such feature is the sand barrier, which is built up usually just above the low-tide line by sediments brought from deep water by waves and longshore currents. The development of barriers is favored by the availability of large sand supplies, such as from a nearby river delta or deposits of glacial drift. The barriers may become long, permanent islands which, as along the coasts of Long Island, New Jersey, and parts of eastern Florida, are much developed for recreational and other uses. Frequently they are connected by bridges to the mainland. They tend to grow seaward during periods of quiet water, but may be breached and reshaped nearer shore during stormy seasons.

A barrier island: Island Beach is one stretch of an almost continuous sand barrier stretching over 100 miles down the New Jersey coast. Most of the sand has been supplied by streams.

The great Cape Cod spit at Provincetown: Currents generated by Atlantic waves beating against the eastern shore carry sediments to the Cape's extremities. There the sediments are dropped as the currents turn the corner. Note the tendency of protected Provincetown Bay (*center*) to fill.

A tombolo on the California coast: In Morro Bay, waves and currents built the tombolo to the left of the island. Jetties on the other side help to maintain a channel. Note building up of the beach at right.

Sand barriers that have become naturally connected to the land at one end are known as spits. These long, curving ridges trail from the point of a headland that has blocked longshore currents and caused them to deposit their sediments. Once started, the building of the spit can be accelerated by beach drift from windward as well as by the gain of additional current-borne sediments. Since barrier islands are themselves obstacles, spits may be found at their extremities. Spits often curve into a hooklike form, like that of Sandy Hook, New Jersey, or the north and south ends of Cape Cod, or England's Spurn Head in the North Sea. Sometimes a series of hooks develops along a spit. If a spit grows across the front of a bay, it becomes what is known as a baymouth bar.

With time the lagoon between a barrier island and the mainland may fill with sediments washed over the barrier during storms, imported by longshore currents and beach drift, and blown or washed in from the mainland. The lagoon may connect with the sea only by way of tidal inlets through the barrier or by ground-water seepage. Lagoons often evolve into salt marshes and mud flats which receive sea water only at high tide. Continued filling or draining by man completes the conversion of a lagoon into dry land.

Shores with barriers may also have what geomorphologists elegantly call cuspate forelands. These are numerous from North Carolina southward along Florida and westward to the Mississippi Delta. Some of these cusp- or wedge-shaped sedimentary features build out at points where obstacles block the normal coastwise movement of sediments; others apparently take shape at eddy points between currents where they turn seaward. Cape Hatteras is a cuspate foreland believed to have been formed by eddies of the Gulf Stream.

Where a small offshore island blocks shore drift, a cuspate foreland may grow out to it. Such a tie is called a tombolo. Occasionally an island acquires two tombolos, with a lagoon between them. If the lagoon fills in completely, as happened at Gibraltar, the island becomes a part of the mainland.

Human Interference

The ceaseless movement of sediments presents vexing problems to beach and harbor engineers. Where the bottom slope of a shore is gentle, channels and harbors silt up, and bars or other depositional features form where they are not wanted. Where the slope is steep, the problem is likely to be the loss of sediments that are wanted.

If a channel in a harbor is deepened by dredging, natural refilling usually follows. Currents working over the harbor bottom leave sediments in the channel just as a broom leaves dust in a crack in a floor. Dredging

A shore rich in sand: The tide marshes of Cape Hatteras, North Carolina, have been built mainly of erosion sediments originating inland.

done once probably will have to be done again and again. Likewise, if sand is dumped into water in order to build out a beach, waves and currents may tend to restore the original shore line by picking up the added sand and distributing it downshore or carrying it out into deep water.

If basic conditions along a shore are changed, however, the restoration of the old balance cannot occur, and a new shore equilibrium with new features will evolve. Thus when a jetty is built to widen a beach, the sand that is trapped on the windward side is subtracted from the supply available for the beach on the leeward side. After this beach has been narrowed as usual by winter storms, it cannot build out again in the spring. If its indignant owner then builds a jetty himself, his beach may recover, but at the expense of the next one down the line. So the chain may be continued for miles.

Such beach problems have been common along coasts of southern California and Florida that are short on sand. Here the best preventive against damage to a beach downwind from a jetty has been the presence of a rocky headland just below the jetty; such an obstacle resists erosion and tends to break the chain. Sometimes, too, a series of groins—short, jutting concrete walls—will serve to spread out the erosion effects downshore.

Some coasts that are being eroded severely appear to be beyond sav-

Harbor strategy at Atlantic City: At this point on the New Jersey shore, groins (*lower left*) are preventing longshore movement of sand. Beaches have been created by means of jetties (*extreme lower left and low center*). Tidal channels are visible at upper right. Note wave refraction outside the narrows.

ing. The eastern face of Cape Cod is exposed to the full force of Atlantic storms, and since the Cape is composed of sand and clay without bedrock ramparts, it is being cut back at the rapid rate of over three feet per year. Of the enormous volumes of sediments washed out of the soft cliffs, especially during winter storms, some are carried seaward by rip currents; much more is moved north and south by the longshore currents, which are constantly extending the long, recurved spits at the tips of the Cape. A bold but vulnerable prominence such as Cape Cod cannot last long against the sea.

The Yorkshire coast of England is not a promontory but, like Cape Cod, it consists of weak glacial deposits and has suffered rapid erosion. Since Roman times, North Sea storms have cut back the shore as much as 3 miles along a 35-mile stretch between Spurn Head and Flamborough, wiping out the sites of several Roman villages. During the past century the rate of shore destruction has been 5 to 6 feet per year.

Special Cases: All Shores

It seems evident that marine erosion, reinforced by subaerial processes, can eventually transform a high, cliffy shore facing deep water into a low shore fringed with masses of loose sediments and facing shallow water. Erosion of the low shore diminishes as the wave-cut bench widens seaward and the waves, breaking farther and farther out, reach the shore with less and less force. An excess of sediments may be carried to leeward along shore and down the continental slope, so that they present lessening resistance to waves and currents. The logical end of these processes could be a shore reduced virtually to sea level, with a wave-cut bench that permitted minimum wave and current attack on the shore, and a rate of sediment production exactly in balance with the rate of removal. Thus a seashore erosion cycle might be completed—setting the stage for a new uplift and a new cycle.

A few decades ago it was thought that many shores have gone through such a cycle. But extensive observation has more recently raised doubts. Any cycle that may have begun a million years ago on the shores of northeastern North America, for example, has certainly been thrown out of gear by the dumping of Pleistocene glacial drift. Along shores where volcanoes have erupted during recent millennia, flows and falls of lava have obliterated profiles due to marine erosion. Evolution cannot occur uninterruptedly on diastrophically active coasts. Shore building by stream deposition on some coasts proceeds faster than erosion. The post-Pleistocene rise in sea level—a rise possibly as great as three hundred feet—has certainly affected shore evolution everywhere. Finally, the edges of the continents show few wide wave-cut benches such as the Strandflat north

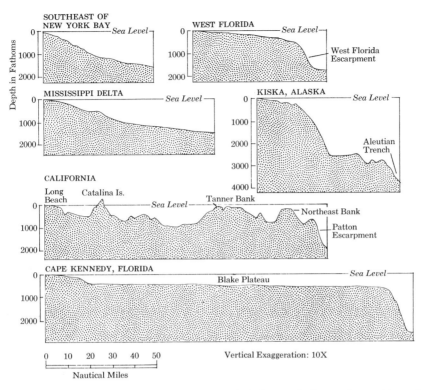

Edges of a continent: Profiles of the continental shelves and slopes show wide variations, reflecting widely differing origins. Wave-cut benches are uncommon.

of Bergen, Norway, which has often been cited as an example of the last stage of a shore cycle.

Coastal experts do generally agree that every shore, whatever its origin and present state, is evolving toward a profile that would present the minimum resistance to marine erosion. The upward curve of the sea bottom toward land is such that waves break and currents move with the least possible erosive effect. The exact curve will vary from shore to shore, according to the supply of sediments, the strength and direction of prevailing winds, and other local conditions. Few if any shores, however, are likely ever to reach a state of perfect equilibrium. If any shore is nearing it today, some event—diastrophic, volcanic, or otherwise—will probably alter the situation very soon.

Because shores are of such diverse origins and have been subjected to so many upsetting events, the "standard" or "model" shore usually corresponds rather poorly to real shores. What has been happening on these real shores depends largely upon the local conditions. Some of these conditions are rather common, and they can be seen again and again.

ROCK, TIME, AND LANDFORMS · 318

The notable aspect of many coasts, especially as seen on a map, is their irregularity. This is due largely to the recent rise in sea level, which converted coastal highland into islands, and river mouths into estuaries. Estuaries are seen along most coasts—the most scenic forms being fiords—and they account for the majority of natural harbors.

Relief maps demonstrate how bedrock structures may control coastline

The Dover Cliffs: The extreme steepness of these cliffs on the English Channel is due to rapid undercutting of the weak chalk by storm waves.

The uplifted coast of the Emerald Bay region in California: The flat surfaces on the headlands south of Los Angeles are parts of an old wave-cut platform. At their present level, waves and currents have created beaches between the headlands and have dissected some headlands to make stacks.

contours. Where the strike of sedimentary rocks is nearly parallel to the shore, deep indentations are rare, and islands formed by the sea-level rise tend to be elongated parallel to the shore. So it is along the Adriatic coast of Yugoslavia and the American Pacific coasts. Such coasts are said to be of the "Pacific" type. Deep embayments that do exist, such as that at San Francisco, are likely to be associated with faulting.

Where the strike is more nearly perpendicular to the shore, embayments are deeper and more numerous, and offshore islands are elongated perpendicular to the coastline. Such shores are noticed in northeastern North America, Norway, the Brittany coast of France, northern Scotland,

and the northern edge of the Mediterranean Sea. These are all coasts of the so-called ria or "Atlantic" type.

In higher latitudes where Pleistocene valley glaciation was intense, both Pacific- and Atlantic-type coasts offer spectacular scenery. Fiords increase as one moves north from about 50 degrees north latitude and south from about 40 degrees south latitude. The fiord coast of Norway is interesting to compare with the northern coast of Spain because both show a rock structure with a strike nearly perpendicular to the shoreline. The Spanish coast lacks, of course, the deep glacial gouging.

Some coasts show strongly the influence of specific rock types. Seas around Scotland, eastern Ireland, and Iceland beat against towering basalt columns and fallen blocks, as at Giant's Causeway and Fingal's Cave. The storied white cliffs of Dover and the opposite French coast are of chalk, which is being kept steep and spectacular by very rapid undercutting—at the rate of one to three feet per year. The granite cliffs of Cornwall, at England's southern end, and the cliffs of slate and quartzite on Brittany's shores, in France, may last longer even though exposed to fiercer wave attack. Italy, Yugoslavia, and Greece have high, rocky shores many of which are of limestone and thus exhibit, as at Capri, an abundance of grottoes and other solution features.

In the eastern Mediterranean, from Syria to eastern Egypt, the shore is astonishingly straight—almost harborless and islandless. Here the coast is of recently uplifted, nearly horizontal strata on which streams have not yet accomplished deep cutting. A comparable coastline is seen in South America north and south of Buenos Aires. Another low coast, but one that has sunk and become deeply embayed, is that of the Chesapeake Bay region—a ria coast.

Along low Atlantic coasts in middle latitudes, most of the fringing islands are depositional, having been built up where the sand supply is abundant. From New York southward and from southern Brazil northward, islands of inundation are few. But along the northern fringes of North America and Eurasia the contrary is the case. These coasts are still depressed from the weight of the Pleistocene ice; they have numerous large, low islands of inundation with broad shallows around them. As the coasts continue to rebound, some islands will become part of the mainland.

Florida's eastern shore represents a seeming anomaly. Its bedrock is limestone, which is weak against wave and current attack; yet limestone debris is lacking along much of Florida's Atlantic seaboard and shore erosion is a problem. Erosion debris is carried away northward and southward as fast as it is produced; hence eastern Florida has a relatively straight shore with long, narrow sand barriers.

The western shore of Florida has no such problem, because it belongs to the Gulf Coast. The Gulf is richly supplied with sediments from several great rivers, including the Chattahoochee, the Mississippi, and the Rio Grande. These materials become trapped in the Gulf because of its shape. The Gulf Coast, then, is a showplace for depositional features, from the southern tip of Florida to the northeastern coast of Mexico.

Where rivers descend to the sea steeply, the sediments are scattered widely and only the mouths of the valleys remain as estuaries. So it is with the ria coast of Spain. Along low coasts, especially strongly indented ones, the rivers can build sizable deltas. Along the Gulf Coast even the violent storms do not sweep the sediments away, and deltas wax to great dimensions. The Mississippi and the Amazon, the Ganges and the Yangtse, the Nile and the Rhine—all these have been able to deliver sediments to the shore faster than waves and currents can carry them away.

Seashores sculptured in glacial debris are found along low coasts where the edges of the Pleistocene ice sheets decayed. Cape Cod and Long Island consist of large moraines and masses of glacial outwash dropped at the southernmost limits of lobes of the Laurentide ice sheet. In the harbor of Boston, Massachusetts, a cluster of islands with rounded tops and often elliptical outlines has been formed by the inundation of drumlins.

On some coasts the diastrophic influence is obvious. Florida, for example, consists of a recently raised anticline; its shores are the limbs of the fold. The California coast from end to end has been heaved up with the Coast Ranges, and wave-cut terraces 1400 feet above sea level near Ventura show how rapid this uplift has been within recent time. Wave-cut

A retreating shore: Storm waves periodically undercut the weak sand and clay cliffs of Cape Cod. Gravity pulls down material loosened by rainwash. In most years the cliffs retreat several feet before the assaults of the sea.

terraces are seen all along the Pacific rim, from Chile to the Aleutians and through the Japanese islands down to the southwest Pacific. Downfaulting, on the other hand, has created the great graben which is now occupied by the Gulf of California, and also the basins of the Red Sea and the Persian Gulf.

The volcanic island arcs of both the Atlantic and the Pacific are young and active, and many are growing by volcanism or uplift. Their coasts tend to be rocky and cliffy. The beaches of coral and basalt sand are usually narrow, and bars and sand barriers are unusual, though some of these islands are ringed with reefs of coral and algae.

The Hawaiian Islands, although they are domes, present rugged fronts. Their basalt is so vulnerable that the shores can but weakly resist the powerful trade winds; the eastern shores are being cut back fastest. Steep cliffs are common, especially along coasts unshielded by neighboring islands. Wave action has reduced Molokai to less than half of its original size, and Nihau is at least half gone. Sand, consisting mostly of olivine and shell or coral debris, is produced rapidly but is insufficient for the construction of depositional features, because most of it drifts down the sides of the domes toward the miles-deep ocean floor.

Lakeshore Miniatures

Lakeshores show seashore features on a small scale. Even the largest lakes have tides of but slight range—Lake Erie's, for example, is less than

A cliffy shore in Hawaii: The land at Sea Arch Overlook, as elsewhere in Hawaii, was built up by lava flows. The vulnerability of the closely jointed, sometimes porous basaltic rock is apparent.

The shore of Mackinac Island in Lake Michigan: Sand is lacking here, and limestone cobbles form the narrow, wave-built beach.

an inch—and therefore currents produced by tidal movement are negligible. Mainly because the fetch of waves on lakes is short, lake waves never begin to rival ocean waves in height, and their impact upon the shore is correspondingly mild. Since wave size and speed determine the power of longshore currents, these currents and the rip currents of lakes, also, are relatively weak in destructive power and carrying capacity. Few lakes are old enough to show major destructional features such as coves, stacks, and wave-cut rock benches. Much more common are the various depositional forms, such as spits, bars, and beach ridges.

Around many old natural lakes, above the present shorelines, appear the remains of wave-cut terraces. Whereas the ocean level has risen since the Pleistocene, lake levels have generally gone down, because of the dry post-Pleistocene trend and because of downcutting in lake outlets. The wave-cut terraces of old Lake Bonneville, some of them 1000 feet above the much shrunken Great Salt Lake of our day, come to mind as classic examples of fossil lakeshores.

Lake Bonneville was only one of many. The coasts of the Great Lakes also exhibit higher terraces—terraces that have yielded a detailed history of ice advances and retreats in this region. Clusters of lesser glacial lakes scattered across North America and the northern part of Eurasia also show souvenirs of the old times of higher water. Some of the most interesting terraces are those of lakes that, like giant Lake Agassiz in central North America, fell victim to evaporation and outlet-downcutting, and today no longer exist.

A coral atoll: Swain's Island in the Southwest Pacific as photographed from the air. The scale is approximately 1:20,000. Note the refraction of waves by the island.

XIV

Continents, Islands, and Sea Bottoms

The sea bottoms account for nearly three fourths of Earth's surface, but being invisible to ordinary viewing and difficult to explore even with ingenious scientific equipment, they are still not well known. The sea's unresting surface looks much the same everywhere, and the landless horizon has no shapes but clouds. The outlines of ocean basins cannot be seen as one sees a mountain range or a river valley; they can be comprehended only by resort to a map or a globe. Nor are there any vistas of sea bottoms like those on land. In the shallows of clear water, as in the West Indies, a diver's eye can follow the contours of a coral reef and detect bright fish a few score yards away, but there is no real view of submarine topography, and in all oceans a depth greater than several dozen fathoms brings one to utter darkness. Here the eye and the camera are limited to close-ups; the panoramas are on maps, charts, and globes.

Curiosity about the ocean bottoms dates far back. Ancient philosophers, impressed with the sea's grandeur, supposed that it lay beneath the

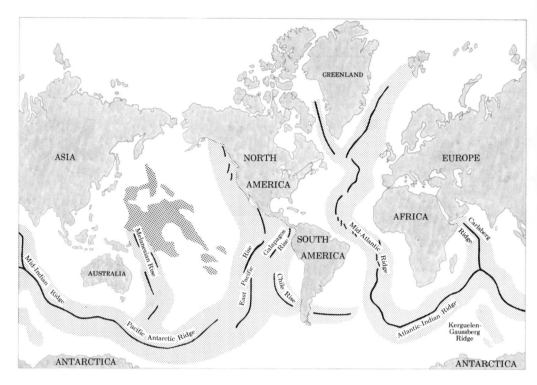

Major relief of the ocean bottoms: The Mid-Ocean Ridge and Rift dominate the scene.

land and that the land was a growth from it. Herodotus and Xanthos of Sardis about 500 B.C. observed that certain northern Mediterranean seaside towns had sunk to become part of the sea bottom and that, on the other hand, some highlands far from the coast contained fossils of marine shellfish. Such observations could have suggested that the contours of sea bottoms are like those of land, and that many terrains were once sea bottoms. Folk literature told of foundered lands such as Atlantis, once known to travelers west of the Gates of Hercules. For practical navigation, Old World mariners with weighted line traced offshore sea bottoms to depths of several hundred feet. Yet in the time of Columbus the Atlantic was popularly regarded as being a bottomless realm of monsters that gulped down ships.

Until a century ago the relief of the ocean bottom beyond the continental shelves was unknown. Shelf contours, particularly off Europe and eastern North America, had been laboriously traced by decades of navigational sounding. But not until the voyage of H.M.S. *Challenger,* the British scientific vessel that traversed the oceans during the period 1872-76, under Sir Wyville Thomson, were the deeps widely sounded. Further surveys were prompted by problems of fishery research and the laying of submarine cables, by increasing scientific curiosity about deep-water organisms, and by the interest of geologists. By 1910 all major types of bottom relief forms that oceanographers recognize today—notably the continental shelves and slopes, the abyssal plains and mid-ocean ridges, submarine seamounts, valleys, and trenches—had been detected, and some of the broad topographic patterns were being traced.

The old-style weighted hand-line—first of hemp rope, then of tough twine, and finally of piano wire—was superseded after World War I by echo-sounding and automatic depth-recording. Plotting of bottom contours was dramatically simplified and speeded up. With the founding of centers such as the Scripps Institution of Oceanography in California and the Lamont Geological Observatory in New York, equipment and techniques were developed for bottom sampling by coring and dredging, direct observation of shelf topography by divers using SCUBA (self-contained underwater breathing apparatus), deep-water photography by means of the bathyscaphe, and seismic sounding—that is, sounding by means of explosion-produced shock waves that are differentially refracted within, and

Continental and sea-bottom profiles compared: These profiles of North America and the Atlantic and Pacific bottoms show a general similarity in the strength of their relief.

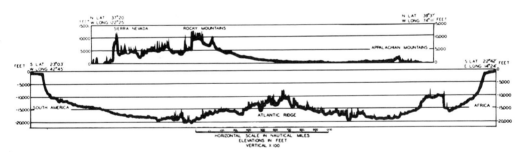

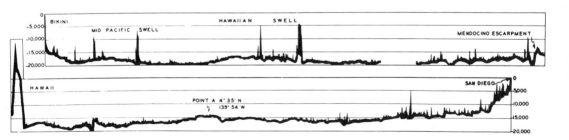

echoed from, rock strata beneath the sea bottom. By midcentury the vast blank areas of the old hydrographic charts had been at least roughly sketched in. As invariably happens with refinement of observation, an imposing array of new questions were raised, yet today the sea bottoms are no longer a "mystery."

Contrary to ideas current until recently, the bedrock of all ocean bottoms is basaltic, though patched with sandy, clayey, or limy sediments. The ocean floors are lower than the continental surfaces because, as we know, the continents are slightly lighter. The basins of the great oceans and of the fringing seas as well—the Mediterranean, the Black, the Arctic, and the rest—appear to be isostatic in nature. In the ocean basins, as on the continents, diastrophism has combined with volcanism and erosion to produce relief of wider diversity and impressive dimensions.

The continents generally are fringed with gently sloping, shelflike platforms, often moderately irregular in relief. At their seaward edge most of these platforms increase in dip and become the continental slope, which runs scores of miles ordinarily (and hundreds rarely) down to the deep ocean floor. The edges of some shelves are fault scarps with steep faces. The slopes often are interrupted by scarps and plateaus, and may ease in gradient to form a continental rise before reaching the floor of the ocean basin.

This floor, once thought to be flat, is now known to have relief much like that of the continents. The broad, gently rolling abyssal plains suggest prairie lands. Submarine mountain ranges are miles high, and a few of their summits form mid-ocean islands. The sea floor has numerous volcanoes. The sea swells are its domes and horsts, the much-discussed trenches are its grabens, and some continental slopes are fault scarps ranking with the world's greatest. Even in erosion the sea bottoms are kin to our landscapes. Beneath the waves there are rushing rivers, the density and turbidity currents, which compare in vigor with their terrestrial cousins.

The Shelves: Submerged Edges of the Continents

The continental shelves, fringing nearly all the world's coasts, usually slope seaward about 10 feet per mile and average 42 miles in width, but spans as short as a few miles and as long as 100 are common. Shelf depths average 72 fathoms but range from a few fathoms to 100. Most shelves represent the top surfaces of masses of sediments, more or less consolidated, that have been eroded off the continents. Some are platforms of very ancient, perhaps original continental rock. A number of shelves appear to be classic wave-cut benches with wave-built terraces along their outer margins.

Portions of the shelves were dry land during the Pleistocene, when so much of the world's water was locked up in glacial ice. The post-Pleistocene rise in sea level has submerged these lands with their river-valley systems—systems that can be traced today on the shelf between Maine and Delaware, for example, and in the English Channel on the portion of the shelf known as the Hurd Deep. Various other shelves appear to have been submerged since very ancient geologic time. Some of them, like those around the Gulf of Mexico, are sinking under a continuously growing load of delta sediments. Shelves in arctic regions are still rebounding after Pleistocene ice-loading, and parts of them—in the Hudson Bay region, for example—will become dry land within the next few thousand years.

Shelves vary so widely that "types" can be recognized only with difficulty, but a few salient differences are worth noting. Off coasts that flank young mountain ranges, such as the Pacific Coast of the Americas generally, the continental shelf tends to be narrow with strong relief. The steep gradient to the ocean floor favors rapid erosion of shelf and slope by the sediment-loaded bottom flows known as turbidity currents. Sediments on these steep slopes tend to gravitate to the ocean floor. In contrast are the shelves along older, lower-lying coasts, such as those off eastern North America. Many of these, like the platform off New Jersey, are relatively flat, with contours like the relief of the coastal plain. Sand, mud, and gravel are abundant and irregularly distributed, with the sand forming numerous low ridges.

Shelves that were glaciated during the Pleistocene are dotted with ice-gouged basins and with low hills and ridges shaped from glacial drift. Georges Bank, the Grand Banks, and other famous fishing areas east and northeast of New England are shoals of glacial debris that was dumped at the edges of ice sheets. The sediments here, including many cobbles and boulders, account in part for the good fishing: they offer hiding places to the small fish, and because the small fish come the big ones come, too. Other glaciated shelves, such as those off Norway and the western coast of southern South America, are cut by deep canyons which are the submarine extensions of fiords.

Where shelves are swept by strong currents, as the shelf off eastern Florida is swept by the Gulf Stream, the sediment layer is thin, most of this material having been carried away or pushed down the continental slopes. The narrowness of the shelf off eastern Florida is probably due mostly to faulting but may have been influenced also by the Gulf Stream's erosional activities. This famous current, with a speed of about three knots at depth and as much as six at the surface, may perform tasks of erosion comparable to those of rivers on land.

The wide platforms around the Gulf of Mexico, reaching widths of more than one hundred miles west of Florida, are sinking under the loads of sediments being poured onto them by the Mississippi, the Rio Grande,

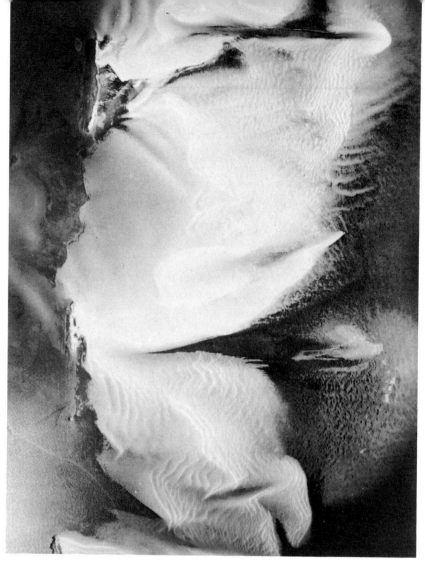

On the bottom: Underwater "dunes" on Bahama Banks, northwest of Great Exuma, are revealed by aerial photography. Motions of water at depth mold the sands.

the Mobile, and other rivers. The magnitude of the sinking is indicated by the fact that the Mississippi Delta reaches a thickness of thirty thousand feet. The Gulf shelves are notable for their salt domes and have been much drilled for oil. The sediments are mostly of mud, with organic refuse mixed in; a zone of sand begins at the outer edges of the mud. Terraces in some of the sediments are evidence of interruptions in the erosion cycles of the contributing rivers.

Around the world generally, the sediments on most shelves vary widely. Rock waste from land and sea-bottom erosion usually predominates. Shell debris and other organic wastes are abundant especially in warmer waters. The shelf off eastern Australia is the prime example of platforms that are heavily loaded with coral. Added to the potpourri of rock waste and organic remains are chemical precipitates from the sea water.

Off the edges of a delta, sediments usually get finer as one moves farther from land, because river waters entering the ocean drop the heavier-caliber sediments first. But off most coasts the longshore currents, rip currents, and even bottom currents miles out keep mixing up the sediments. Mud accumulations can be thick near shore, and sand and gravel are often brought up by research vessels from the outer edges of shelves and even from the continental slopes. The uneven distribution of shelf sediments in some localities is due not only to the work of currents but also to erosion of the shelves during past epochs when they were dry land.

Continental Slopes: Fringing the Deeps

All the continents have distinct continental slopes even where shelves are poorly developed or missing. The line where the shelf ends and the slope begins is usually marked by a moderate increase in steepness. Around the margins of the Pacific, the gradient of the slope is rather consistent down to the deeps at 3000 to 4000 fathoms, averaging a little over 5 degrees. The total drop from the summits of the Andes Mountains of South America to the sea floor is some 43,000 feet, and this drop occurs with slight interruption. Off the mouths of rivers such as the Amazon and the Mississippi, however, the grades are more gradual, around 1 to 2 degrees, and the slopes are often scores of miles in width. Here the accumulations of sediments strongly influence slope contours.

Before reaching the deeps, ordinarily at 1500 to 2000 fathoms, continental slopes of the gentler variety sometimes grade into a less steeply sloping terrace called a plateau. Such plateaus may be scores or even hundreds of miles wide. The Blake Plateau, extending from South Carolina to the Bahamas, reaches a width of about 150 miles and a depth of 400 to 600 fathoms. Its eastern edge is a scarp that plunges with a gradient of as much as 45 degrees down to the ocean floor, at about 1800 fathoms. Plateaus such as the Blake tend to be relatively flat.

Most continental slopes, especially the steeper ones, show strong relief in the form of valleys and ridges. These are generally perpendicular to the coastline, but dissection is usually less advanced on continental slopes than on the flanks of terrestrial highlands.

Sand and gravel of land origin are found on slopes out to one thousand fathoms and more, but mud and ooze of marine origin, with shells and other organic debris, predominate. Mud is abundant on slopes near the mouths of great rivers. Some slopes, especially the steeper ones, are rocky and relatively bare and are kept irregular by frequent slides.

Many slopes show evidence of strong intermittent faulting. Frequently steplike profiles suggest the patterns of step faulting familiar on land. Although some valleys in the slopes are winding canyons made by turbidity currents or are the winding seaward extensions of drowned river valleys, many others such as those in the Aleutian slopes are revealed by soundings to have the typical angular contours of fault valleys.

At the foot of at least half of the continental slopes so far investigated, particularly around the Pacific, deep trenches have been identified. Seismic sounding off the eastern coast of the United States indicates the existence here of old trenches that have become filled. The younger trenches of the Pacific and Caribbean, especially, are centers of the most intense earthquake activity. Here apparently are the active faults between the continental blocks and the ocean-basin blocks, where readjustments are continual as sediments eroded off the continents keep loading the fringes of the ocean basins.

Some slopes correspond at least roughly, as do some continental shelves, to the adjacent coasts. The slopes off the northeastern United States, Spain, and the Mediterranean countries are much dissected like the coasts. The steep slope off the Andes likewise corresponds to the coast, canyons being numerous in the south where the coast is indented with fiords. By contrast, the slope between Cape Hatteras and Cape Kennedy on the Atlantic Coast shows little continuity with the land. Here the rise seen in the slope to the north disappears, and instead there is the Blake Plateau with its precipitous outer fault scarp.

The slope off western Florida's broad shelf is easy to about 600 fathoms, then drops sharply along a fault scarp with inclinations as great as 27 degrees—nearly matching the slope off southeast Cuba, which is the world's steepest. West of the Mississippi Delta the landward part of the slope is inclined only about 1 degree, and like the coastal plain is marked by basins, valleys, and hills. (Some of the hills may be salt domes.) This area ends at a steep escarpment, five hundred fathoms in height, beneath which the slope extends to the Sigsbee Deep at two thousand fathoms.

The Pacific slopes, generally steeper than those of the Atlantic, suggest more recent and more intense crustal activity. In steepness and in topography they are bewilderingly diverse. The slopes from Cape Horn to Vancouver average around 5 degrees, but toward the north they steepen. Canyons of erosion are met intermittently from southern California to Vancouver; then fault troughs predominate around the northern Pacific arc as far as Japan. Down the western side of the Pacific the slopes though highly variable are characteristically steep, reaching inclinations of 25 degrees north of Kamchatka, 2 to 10 degrees off southern Japan, and 10 to 12 degrees or more from the vicinity of the Philippines to Australia.

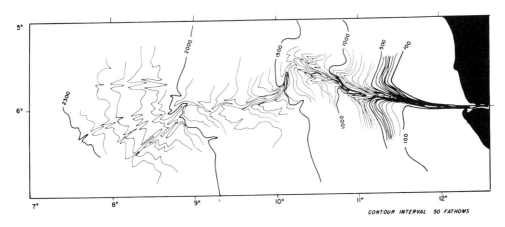

The Congo's underwater canyon: This gorge is an extension of the river valley.

The slope from western Spain and Portugal is only 1 to 3 degrees, increasing to 15 degrees off the ria coast of northern Spain. Northward the pitch eases again, and off western France clusters of canyons appear. Around the British Isles the slopes are gentle, but the bottoms here are sharply irregular, with many scarps and ridges. Far out on the slopes bordering northern Europe, at depths of 1000 fathoms and more, are thick deposits of sand and gravel—material rafted out to sea by Pleistocene icebergs. Off southern Scandinavia the slopes are as gentle as 1 degree, but a steepening to about 5 degrees occurs northward, and submarine canyons grow more numerous. In the Arctic north of Alaska part of the slope represents a submerged portion of the continent, with a steep scarp of 23 degrees at the seaward edge.

The Puzzle of Submarine Canyons

The grabens so common in the Pacific slopes and those of the North Atlantic clearly correspond to the grabens of the land. But the valleys of erosion, properly called submarine canyons, are less easily understood. A few are obviously the seaward ends of drowned river valleys, but most are not, and these present some puzzles.

Submarine extensions of river valleys can be accounted for by either subsidence of coasts or the worldwide (eustatic) rise in sea level after the Pleistocene. Such canyons have already been mentioned as abundant off northern Mediterranean countries, northern Spain, western Africa, southern California, and the middle Atlantic states of North America. Hudson Canyon, which runs about 60 miles southeastward from the estuary of the Hudson River (New York harbor), cutting 4000 feet into the continental slope and eventually fading out at a depth of 14,000 feet, is one of the largest of these land-sea canyons. Among the series off southern California the largest is Monterey Canyon, which reaches 30 miles down the slope and has walls rivaling those of Grand Canyon. The longest specimen in the world is the one that begins at the mouth of Africa's Congo River and runs 150 miles westward over the Atlantic bottom.

Most submarine canyons have no landward extensions in the form of river valleys. Many do not even extend up into the shelf, but have their heads in the continental slopes; canyons of this type have been detected off the New Jersey coast, southern California, and Tokyo harbor, to mention just a few. Some canyons, especially those that have tributaries, could be remnants of very old river valleys whose landward portions have been obliterated by erosion or covered with sediments. But canyons that head in the continental slope must be largely the work of turbidity currents. These are currents which, because of debris held in suspension, sink below lighter adjacent waters and, in so doing, build up considerable speed and erosive capability.

Hudson Canyon, although it is of the land-sea variety, illustrates the

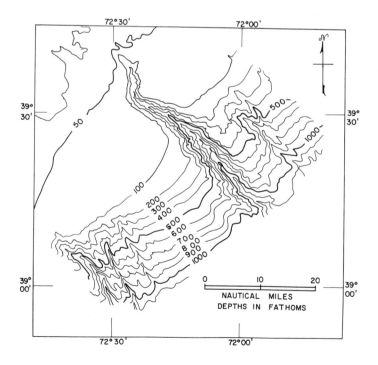

Hudson Canyon: To what extent the canyon represents subaerial erosion and to what extent submarine erosion remains a subject of controversy.

problem. All investigators agree that its upper reaches, at least to a depth of a few hundred feet, were originally cut by rivers, and later by glacial ice and meltwater during the low-water stages of the Pleistocene. But what of the several thousand feet of vertical cutting that has occurred below the Pleistocene sea levels, which cannot have been more than a few hundred feet lower than those of today? Turbidity currents are offered as the explanation. Current speeds of 50 miles per hour, far greater than the speeds achieved by rivers on land, have been proposed as likely for turbidity currents on such a steep, long slope. If turbidity currents are not accountable, a river must be. But a river would need a million years or more to cut through 4000 feet of bedrock, and then the land would have to sink that much in order to drown the canyon. Such deep sinking of a coast would be extraordinary.

Yet there are counterarguments. For one, the sediments being carried by turbidity currents down through the present canyons are mostly light and fine, more typical of gentle currents than of the raging torrents that ordinarily do deep canyon cutting. Again, if mountains can be raised up several thousand feet, perhaps seacoasts can subside that much. Finally, there is the possibility that the cutting of Hudson Canyon began so far back in geologic time that gentle currents have been able to accomplish the deep cutting and a coastal subsidence of several thousand feet is believable. Considering all the arguments, we can be sure only that the dispute will continue until the ocean bottoms have been much more thoroughly sounded, sampled, and interpreted than they have been to date.

On the Bottom: Abyssal Plains and Trenches

The broad, relatively flat expanses below the continental slopes and rises are known as abyssal plains. These correspond to the broad, high plains of the continents. The abyssal plains are covered with deep-sea oozes of organic and chemical origin, mixed with reddish-brown clays consisting of sea-water precipitates and oxidized meteorite fragments accumulated over millions of years, and also gravels, sands, and muds transported here by winds, turbidity currents, and melting icebergs. Glacial sediments litter the bottoms of the north Atlantic particularly.

The abyssal plains have little relief. Their irregularities have been modified by the accumulation of sediments, and since these are more plentiful near the continents, it is near the continents that the plains are smoothest. For the same reason, the abyssal plains toward the center of the Pacific are more irregular in relief than those of the Atlantic, which is a smaller ocean.

Abyssal plains show here and there features of both erosional and diastrophic origin. From the foot of some canyons in continental slopes great fans of sediment have spread, and through them channels run out onto the plains. Often these channels are flanked by natural levees suggesting those along the lower Mississippi. Out on the plains channels have been cut through the sides of shallow basins and low rises: these certainly are the work of turbidity currents. Evidences of faulting on the abyssal plains include sea scarps and the deep-sea trenches.

The trenches have inspired much interest as sites of the deepest ocean soundings and the epicenters of the world's strongest earthquakes. They occur mostly in series along the volcanic island arcs bordering the Pacific Ocean, notably the Aleutians, Kuriles, the Marianas, and along the Lesser Antilles in the Caribbean. Typically they have depths of over 4000 fathoms and lengths of hundreds of miles. Walls suggesting a V-shape are common, but most have flat bottoms—the flatness being due in some instances, at least, to filling with sediments. The deepest trenches are those of the western Pacific, ranging from Kamchatka to New Zealand. Ten of them have floors deeper than 4200 fathoms, the record depth being that of the Challenger (Trieste) Deep in the Mariana Trench near Guam: 5933 fathoms. The latter's closest rival is the Tonga Trench, with a bottom at 5905 fathoms.

In the Caribbean the Puerto Rico trench, 600 miles long, plunges to 5030 fathoms. Recent soundings indicate that the walls of this flat-floored trough are stepped, perhaps like the step-faulted slopes on land topography. Shallower, shorter trenches are encountered elsewhere in the Atlantic along the curve of the Sandwich Islands and east of the Mid-Atlantic Ridge near the equator.

Gravity measurements reveal that the crust under the trenches is deficient in mass. The crust would stand higher if isostatic balance were being maintained. Apparently the trenches are being dragged down by subcrustal forces, perhaps convectional currents, or forced down by lateral thrusting forces, contrary to normal balancing tendencies of the crust. The fact that ocean trenches lie in the zone of maximum volcanism and earthquakes is significant. The down-dragging, or down-thrusting, of the crust beneath trenches is confirmed by the down-bending of the Mohorovičić Discontinuity beneath them. This distortion of the "Moho," or borderline between crust and mantle, has been detected by seismic sounding.

Submarine Mountain Ranges

Among all the features of the deep sea bottoms none are more remarkable than the mid-ocean ridges—or, if these are considered as continuous, the Mid-Ocean Ridge. Dominating the Atlantic bottom is the Mid-Atlantic Ridge, a submarine mountain range which winds down the middle of the ocean from Spitsbergen to the latitude of Cape Horn, keeping almost exactly on the midline between the Americas on the west and Africa and

A section of the mid-Iceland rift zone: Here is the Thingvellir graben, part of the Mid-Atlantic Rift. The cliffs are young fault scarps. In the far background, at right, is the famous Skjaldbreidur —Icelandic for "shield volcano"—from which all other shield volcanoes have taken their generic name.

Europe on the east. This range rises about 2000 fathoms above the ocean floor, which in the vicinity averages 3000 fathoms deep, and it is as ruggedly mountainous as any land surfaces, though detailed soundings indicate fewer valleys of erosional origin. Many segments of the ridge are terraced. The highest points emerge above water to form the Azores and the St. Peter and St. Paul Rocks in the North Atlantic, and the islands of Ascension, Tristan da Cunha, and Diego Alvarez in the South Atlantic. The deep-crustal nature of the ridge is all but proved by the typical rock samples dredged up from it: basalt, diabase, gabbro, serpentine. The few fragments of continental-type rock that have been brought up—shale and limestone—are probably ballast dropped from ships of bygone days or rocks of land origin rafted out to sea by icebergs or by masses of kelp.

A rift valley, or chain of grabens, has been traced intermittently along the Mid-Atlantic Ridge. It averages 12 miles in width and drops 900 to 1200 fathoms below the ridge summits. Along the ridge earthquake activity is frequent.

In the South Atlantic, at about the latitude of Cape Town, a branch of the Mid-Atlantic Ridge known as the Rio Grande Ridge runs northward toward the middle eastern coast of South America, and a twin feature called the Walvis Ridge extends northeast to the southwestern coast of

Africa. The Mid-Atlantic Ridge proper continues southward, then continues east to become the Atlantic-Indian Ridge, which in turn leads to a complex of submarine mountain chains branching south, north, and east in the Indian Ocean. The main ridge then runs south of Australia into the south-central Pacific, whence it turns northeast as the East Pacific Rise. About midway between Australia and South America this ridge divides, one arm turning northwest to become a cluster of ridges in the mid-Pacific and northeast of Japan, and the other arm—still as the Pacific Rise—continuing northeast toward Mexico. The elevations of all these mountain chains above the deep ocean floors are comparable to the elevations of the Atlantic ridges. As the accompanying map shows, many island groups of the Pacific are formed by crests of the ridges.

Soundings suggest the presence of rift valleys on some segments of the Pacific and Indian ocean mountain chains. Since fault features are difficult to distinguish positively on rugged submarine topography, further exploration will be necessary to verify them. As in the Atlantic, the ridges mark lines of intense earthquake activity.

A number of great rises exist on the ocean floors. They are broad arches that match the ridges in size but not in ruggedness. The greatest is the East Pacific Rise, already mentioned as reaching from the south-central Pacific to Mexico; it is several thousand miles long and reaches 1000 to 1600 fathoms above the Pacific bottom. Many earthquakes originate here, but no rift valleys have been positively identified. Bisected east to west by many mountain ridges, this rise may be a branch of the mountain system that we know as the Rockies in North America and the Andes in South America.

Oceanic Islands and Seamounts

Many oceanic islands apparently are built of lavas poured out in eruptions of volcanoes or in fissure flows. They are to be distinguished from two kinds of offshore islands: those produced by the erosion or submergence of a seacoast, and those built up by coral or algae on shelves. Most coral islands are, in fact, masses of volcanic rock that have become capped with coral. Some volcanic islands are complete cones; others are remnants of cones or flows. All are in the process of rapid marine erosion.

Since all oceanic islands are by definition islands that rise from deep water—say, 500 fathoms or more—they are in reality mountains of imposing stature. Mauna Loa, the great volcanic dome of Hawaii, rises 17,000 feet from the ocean floor to sea level and then another 13,680 feet into the air. Many an oceanic island that rises only a score of feet above the waves is actually the tip of a mountain two miles or more high. Two miles is the average elevation above the bottom of the shorelines of oceanic islands.

The islands of the Pacific—there are thousands—probably do not outnumber the lava mountains that barely fail to get their tops above water. Many of these peaks, which are called seamounts, stand on the ocean ridges; others rise as scattered, solitary volcanoes from the ocean floor itself. In the Pacific Ocean alone more than 1400 seamounts (defined as submarine peaks with a relief of at least 500 fathoms) had been identified by the year 1963. Seamounts are most abundant in the Pacific, in the central areas especially, but the Atlantic too has its seamount chains, the major ones running along the Mid-Atlantic Ridge and eastward from Georges Bank, off New England, to Bermuda.

The flat-topped variety of seamounts, called tablemounts or guyots, have precipitated much professional argument. The tops look as if they were planed off by waves during some past epoch of lower sea level. But these top surfaces are at depths of hundreds of fathoms, and sea level during the Pleistocene—an epoch of extraordinarily low water in geologic history—was never more than three hundred feet lower than it is now. A complicating fact is that while some guyot chains have accordant flat tops, as might be expected from erosion controlled by sea level, the tops of others vary in elevation by hundreds of feet. There is no clear indication that sea level ever remained fixed long enough for the majority of seamounts to be cut down to a standard elevation.

Most geologists now suppose that the flat tops are indeed the result of

A volcano reawakening: In 1962 a new crater developed in the flank of Mount Kiska, in the Aleutian Islands, and lava issuing from it poured into the sea. Here the land is still being built by volcanism.

A land of congealed lava: Koko Head, at the southeastern tip of Oahu, in the Hawaiian Islands, is a volcano that has been extinct for many centuries. Note the edges of many lava flows exposed in the cliff at left.

marine erosion. Some submergence has apparently occurred with post-Pleistocene melting, and even more by downwarping of the crust beneath the guyots. On land the crust often does sink where supporting magma has been withdrawn from beneath it. On sea bottoms perhaps a similar process has occurred beneath guyots through which large amounts of water, gases, and solids have been erupted.

Organic Reefs

Seamounts account for most of the world's coral atolls and barrier reefs. About a century ago Charles Darwin, whose shrewd observations of flora and fauna in the Southwest Pacific led to his evolutionary theories, speculated that some coral atolls must have been formed by the upward growth of rings of coral around subsiding volcanic islands. Others, he thought, might have resulted from the upgrowth of coral masses from shallow banks near land. He believed that if deep enough borings could be made, most coral islands would be found to have a basement of volcanic

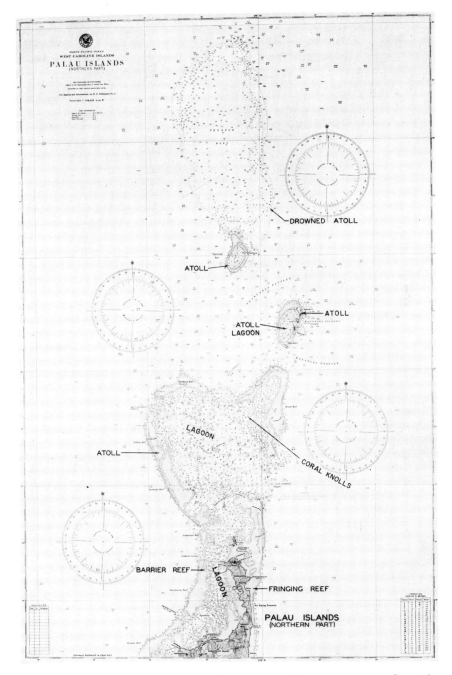

Panorama of coral reefs: On this U.S. Hydrographic Office chart appear the various types of reefs that make up the Palau Islands, in the West Caroline group in the southwest Pacific.

rock. In recent years oceanographers with elaborate modern facilities have established the truth: Darwin was right.

Coral atolls and barrier reefs are the most numerous of the broad class of features known as organic reefs. These consist of masses of the hard body parts of lime-secreting corals and coralline algae; in earlier times numerous reefs were formed also by sponges and bryozoans. Modern corals grow best at less than twenty-five fathoms (the maximum is about ninety) in clear, warm tropical waters. If food supply and temperatures are favorable and the reef-building animals are not harmed by strong influxes of fresh water or mud (as might pour off an island after extraordinarily heavy rain), a reef can grow upward at the rate of ten to twenty fathoms per thousand years. During growth its porous structure tends to fill with organic debris from the corals and algae and from foraminifera; fillings may also include chemical precipitates from the sea water and, happily for modern industry, much later fillings of petroleum. Although growth occurs only underwater, storm waves keep breaking off chunks of coral and piling them up. Where the piles of fragments emerge above water, they form coral islands and barriers, which are then shaped by subaerial and marine erosion.

The first developmental stage for many atolls was the underwater growth of a fringing reef, of coral and coralline algae, on the slopes around an island. Over probably hundreds of millennia the ocean floor beneath the island bent downward, and the island gradually submerged. As subsidence continued, the lower part of the reef died and the higher

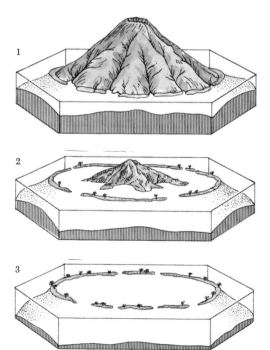

Development of an atoll: As suggested by Charles Darwin, atolls appear to develop as a result of the submergence of a volcano and the upgrowth of a fringing reef around it.

portions kept growing upward, keeping pace with the subsidence. Thus was produced a barrier reef surrounding a lagoon, the latter being approximately centered over the vanished island.

As this reef kept growing, its top was shattered by storm waves, and the heaped-up debris formed the ringlike island that we call an atoll. Storm waves often hurled reef fragments over the barrier into the lagoon, and this rock waste, along with other organic remains and sea-water precipitates, formed a thickening cover on the still-sinking former island. In parts of the lagoon the heaps of debris emerged to form coral knolls—also called pinnacles, or patch reefs. These, with the higher zones of the reef, were attacked by agents of subaerial erosion (those above water), so that sinkholes and caverns were created. During low water in the Pleistocene the upper portions of the reef were exposed to intensive erosion, and later they were submerged again. Thus the atoll of today has developed from the volcanic island of a million or a few million years ago.

The extent of subsidence of such volcanic islands is clearly shown by borings. At Eniwetok Atoll in 1952 two borings reached a volcanic basement of olivine basalt at 4158 and 4610 feet below sea level. Seismic soundings at Bikini and Kwajalein detected a similar basement (evidenced by an abrupt change in the refraction of shock waves) at depths of about 4900 and 5500 feet, respectively. The Eniwetok borings produced fossils of pollen and snails that could have originated only on land at elevations of at least several hundred feet. Such evidence proves beyond a doubt that some if not most Pacific seamounts have sunk thousands of feet.

This evidence of profound subsidence disposes of the theory, once widely held, that atoll formation has resulted from "swinging" sea levels, particularly during and since the Pleistocene. Such fluctuations did leave their marks on the atolls in the form of drowned valleys, but they cannot account for the presence of fossils of land organisms nearly a mile down.

As Darwin supposed, atolls have grown up also from subsiding shallow platforms. In some instances the first stage was the growth of coral and coralline algae over a large area of the platform. During Pleistocene low water, the upper part of the structure was above water and its central area was deepened, irregularly, by solution. Later, rising sea level and continuing subsidence deepened the lagoon even more with respect to the continued upward growth of the surrounding reef. Eniwetok Atoll itself is one of many believed to have developed in this fashion.

The tropical and subtropical seas of the world are adorned by more than three hundred atolls, mostly the work of coralline algae. The majority are in the Southwest Pacific, being most numerous among the Tuamotu Islands, the Carolines and Marshalls, the Fijis, and the islands of eastern Indonesia. The few outside this group include Midway Island, two atolls in the Red Sea, the Dry Tortugas off Florida, and the Hogsty Reef in the Bahamas.

An ancient coral reef in the Southwest: The Guadalupe Mountains of Texas and New Mexico have been carved from a tilted block of coral limestone. El Capitan, shown here, is the highest portion. Farther north, within the block, are the Carlsbad Caverns.

Much more common than atolls are the underwater coral reefs. Generally these flourish best in waters somewhat north or south of the equator, where temperatures are most favorable. Such widely scattered areas as the Hawaiian Islands, the West Indies, and the Persian Gulf all offer colorful underwater realms of coral to the scientist and the diving sportsman.

Geologists recognize in limestone formations on the continents organic reefs of very ancient vintage. Fine details have often been obliterated by filling or by pressure, but the reefs can ordinarily be identified as such by their fossil content and by their steep margins (corresponding to their ancient submarine slopes), lack of stratification, and zones of debris produced by wave erosion. Among the best-known fossil reefs in the United States is the one exposed in the cliffs of El Capitan, the highest mountain in Texas. This has a high content of sponges and algae. In Europe the rock masses that form the Dolomitic Alps have been shown to consist mainly of coral reefs.

Continents versus Sea Bottoms

The concept of continents as blocks that rise above the sea in a balancing process has raised deep questions. How did the continents first become differentiated from the sea bottoms? To what extent have they since changed? What is likely to be their destiny in eons to come?

Until recent years it was generally supposed that the continents began as granitic crusts forming on a cooling molten Earth. This fiery picture of granite masses on a sea of basalt was seized upon by the German geologist Alfred Wegener (1880-1930) for his theory of continental drift, announced in 1924. The continents, said Wegener, were formed as a single original land mass broke up. He was one of many, including Sir Francis Bacon, who had noticed how neatly the continents could be fitted together, notably the eastern coasts of the Americas and the western coasts of Africa and Europe. But his theory never won complete acceptance. Geophysics provided evidence that Earth has been getting warmer, not cooler, since its birth, and apparently the crust never underwent a molten phase. No process could be described that would account for the splitting up of a great land mass and the migration of its parts. Paleontologists, comparing fossils from the various continents, saw little sign that they might have evolved from a single original animal and plant community.

Continental land in the making? The Virgin Islands are part of a volcanic island arc which may eventually become physically part of the American mainland. The view is from Bordeaux Mountain, St. Johns.

Since Wegener's time the crust has been intensively studied by echo and seismic sounding, recordings of heat flow, and the analysis and correlation of continental rocks and ocean sediments. These studies and data on the age and chemical history of the oldest continental rocks have established a few general facts about the continents but leave some basic questions still in the air.

The oldest rocks of each continent are found in the high central areas, or shields, which appear to be the exposed, worn-down nuclei around which the land masses grew. These ancient basement rocks, with radiometric ages as old as 3.6 billion years, plunge to depths as great as seven to nine miles beneath the sedimentaries. Some have been derived from earlier volcanics like those of today's island arcs—arcs which, we are reminded, are the sites of most current volcanic and earthquake activity. As one moves oceanward from the continental shields, the basement rocks grow younger.

It is tempting to conclude that each continent began, perhaps about where it stands today, as an active volcanic ocean ridge rising above the waves of the primordial sea of four billion years ago. Heat produced by radioactive minerals in the crust could have been responsible for this volcanism. As erosion cut down the mountains, the debris accumulated on their flanks and widened the land mass. Outpourings of lava and uplifts due to isostatic readjustments, reaching maxima every few hundred millions of years, offset the degrading effects of erosion, and the continents grew. Today along some continental margins the process of accretion is still going on. The island arcs of our time will tomorrow be part of the main land masses.

Certain other evidence points rather directly to continental drift. Magnetic iron-oxide particles found today in sandstone of the Mesozoic and earlier times have an orientation which suggests that the magnetic poles of those times, and therefore the geographic poles also, were thousands of miles from where they are today. Unless the poles have migrated (and this would imply a change in Earth's axis of rotation), the continents must have moved. The discovery of tillite, the rock formed from glacial till, in late Paleozoic strata in Africa within seven hundred miles of the equator hints either that climate patterns on the planet then differed greatly from today's, with reference to today's poles, or that Africa has drifted.

Oceanographers pondering the rift in the Mid-Atlantic Ridge have considered the possibility that such breaks in the crust are due to the expansion of a warming planet. Could such an expansion have been producing sweeping geographic changes since Earth's earliest days? The rifts do indeed appear to be lines along which the crust is being pulled apart, just as the ocean trenches are zones where the crust is being dragged or squeezed down. The convection theory, according to which slow currents

A changing continental margin: Plateau remnants at the southern tip of Africa were islands for a time but have recently been uplifted to become part of the mainland. The flat-topped highland at the center, rising above Cape Town, is Table Mountain.

of upward- and downward-circulating mantle material are constantly disturbing the crust, seems to fit both of these interpretations of the rifts and trenches.

The deep ocean floors appear to lack sediments more than about 150 million years old—that is, older than the Cretaceous Period. Seismic sounding detects beneath the young sediments only a dense, probably basaltic rock. If this is the original rock of the ocean basins, where are the sediments of the three or four billion years that preceded the Cretaceous? The question implies that the ocean bottoms of today can be no more than 150 million years old, and that continental drift has obliterated previous ones. Could the ocean basins be areas opened up by the drifting apart of continental masses not much more than 150 million years ago? But if the continents really have drifted, why is there no discernible trail of disturbed rocks in their wake? Oceanographic tests indicate that the basement rocks at the eastern edges of the Americas are deeply rooted—and have been since the remote geologic past.

Recently it has been proposed that the ocean basins were made by the impacts of giant meteoroids or asteroids. This possibility has been taken seriously in view of the battering from outer space that has been suffered by the Moon. However, since old sedimentary and metamorphic rocks at the edges of the continents show no strong signs of severe disturbance in recent time, any basin-forming impacts must have occurred very long ago.

An older theory held that the Pacific Ocean occupies a hole left when a mass of material was drawn out of Earth's side by the gravitational attraction of a passing astronomical body, perhaps a comet, or was flung out by our planet's rapid spinning. This mass of material was supposed to have consolidated into spherical form, becoming the Moon. Since the evidence today points to independent formation of the Moon, with later capture by Earth's gravity, a catastrophic explanation of the Pacific basin is unconvincing. Very likely it will turn out that the Pacific basin originated just like the basins of the Atlantic and other oceans. On all ocean bottoms the mantles of sediments, the depths, and the relief are generally similar.

Geologists await eagerly the completion of the Mohole project—or, better, a series of Mohole projects in the several oceans. If it proves practicable to drill through several miles of water and more miles of sediments and other crustal rock to the Mohorovičic Discontinuity, the rock samples that are recovered can be dated and chemically analyzed to determine their histories. This information may tell whether the continents grew in their present positions or, on the contrary, have migrated on the crust and will continue to do so in eons to come.

Certain it is that the ocean bottom holds the answers to profound geologic questions. The sediments and rocks of the deep ocean floor, having been altered far less than the land masses by episodes of uplift and erosion, may yield to future oceanographers an extensive record of Earth's past—a record encompassing the first two billion years that are now so obscure. Submarine events of today, particularly around the edges of the continents, in the trenches, and in the Mid-Ocean Ridge, are part of a sequence that is undoubtedly fundamental in the planet's evolution, past and future. As the story of our seascapes is read with increasing clarity, some of the oldest puzzles about our world may be solved. The seascapes, more than the land, may tell why there are continents and sea bottoms, why there are lands high and low, why the ground shakes and the fire bursts forth, and how the giant hands of change work in the perspective of five billion years. Perhaps they will tell how the world began, and how— long after man's time—it will end.

Further Reading

ADAMS, FRANK DAWSON. *Origin and Development of the Geological Sciences.* Dover Books, New York, 1938 (paperback). An academic but fascinating account of milestones in geology from ancient times to the nineteenth century.

BULLARD, FRED M. *Volcanoes.* University of Texas Press, Austin, 1962. A readable popular account of volcanic phenomena by a volcanologist.

DUNBAR, CARL O. *Historical Geology.* John Wiley and Sons, New York, 1960. A standard text on Earth's history.

DYSON, JAMES L. *The World of Ice.* Alfred A. Knopf, New York, 1962. An absorbing explanation of glaciers and the Ice Age in popular terms.

EARDLEY, A. J. *General College Geology.* Harper & Row, New York, 1965. A basic introductory text.

FENTON AND FENTON. *Giants of Geology.* Doubleday and Co., New York, 1952. Biographies of the pioneer geologists for laymen.

FLINT, RICHARD F. *Glacial and Pleistocene Geology.* John Wiley and Sons, New York, 1957. A specialist's account of glacial phenomena and the Pleistocene Epoch.

HOLMES, ARTHUR. *Principles of Physical Geology.* Ronald Press, New York, 1965.

Unusually well written; contains much foreign material.

LAHEE, F. H. *Field Geology*. McGraw-Hill Book Co., New York, 1961. A handbook which contains much concentrated information regarding the identification of landforms.

LOBECK, A. K. *Geomorphology*. McGraw-Hill Book Co., New York, 1938. An older text still valuable especially for its diagrams and photographs.

MOORE, RAYMOND C. *Historical Geology*. McGraw-Hill Book Co., New York, 1958. A standard text on Earth's history.

MOORE, RUTH. *The Earth We Live On*. Alfred A. Knopf, New York, 1956. A highly readable layman's account of the development of geology as a science.

POUGH, FREDERICK. *Field Guide to the Rocks and Minerals*. Houghton Mifflin Co., 1953. A guide to mineral identification and collecting.

SHEPARD, FRANCIS P. *Submarine Geology*. Harper & Row, New York, 1963. A scholarly summary of current knowledge of coasts, shores, and sea bottoms.

SHIMER, JOHN. *This Sculptured Earth*. Columbia University Press, New York, 1959. Good reading about landforms in the continental United States.

STRAHLER, ARTHUR. *Physical Geography*. John Wiley and Sons, New York, 1951. Includes descriptions of many types of landforms; especially notable for its diagrams.

THORNBURY, WILLIAM D. *Principles of Geomorphology*. John Wiley and Sons, New York, 1954. A carefully reasoned text on landforms and the various theories concerning them.

———. *Regional Geomorphology of the United States*. John Wiley and Sons, New York, 1965. A scholarly, well-illustrated survey of specific localities of the United States, including Alaska and Hawaii.

ZIM, H. S. and SHAFFER, PAUL. *Rocks and Minerals*. Golden Press, New York, 1957. A richly illustrated, easily used identification guide for beginners.

Glossary

Further explanations of most of the following terms, as well as many others, can be located in the text by consulting the index.

ACIDIC ROCK. A group of light-colored igneous rocks, such as granite and rhyolite, in which silica predominates.
AGGRADATION. Building up of the elevation of a surface by deposition of sediments.
ALLUVIAL CONE. A steep-sloped, spreading mass of rock waste deposited by a stream at the foot of a valley.
ALLUVIAL FAN. A wide-spreading, gently sloping, sometimes stratified mass of rock waste deposited by a stream at the foot of a valley.
ALLUVIAL PLAIN. Any terrain made flat by the deposit of sediments by a stream or streams.
ALLUVIUM. Sediments deposited by streams.
ANTICLINE. A fold in which the strata dip in opposite directions from the axis; an upfold.
AVALANCHE. A fast-moving slide of snow or ice on a mountain slope.

GLOSSARY · 352

BADLANDS. An arid terrain deeply and closely gullied and canyoned by streams.

BAJADA. A gradually sloping surface, formed of detritus, extending from alluvial fans or a rock slope toward a bolson.

BASALT. A dark, heavy, fine-grained, extrusive igneous rock consisting about equally of feldspars and ferromagnesian minerals.

BASE LEVEL. The elevation below which a terrain cannot be reduced by stream erosion.

BASIC ROCK. Any of the dark, relatively heavy igneous rocks, such as basalt, that are rich in ferromagnesian minerals.

BEDDING PLANE. In stratified rock, any of the planes that separate the rock layers, or strata.

BEDROCK. The solid rock of Earth's crust.

BLOCK MOUNTAINS. Mountains formed by the tilting or uplift of crustal blocks due to faulting.

BOLSON. A large depression that receives drainage from surrounding highlands but has no outlet.

BOX CANYON. A steep-sided, zigzagging canyon, which gives one the impression of being completely walled in.

BRAIDING. The formation of elongated islands of sediment in a streambed.

BRECCIA. Rock formed by the consolidation of angular rock fragments.

BUTTE. A high, steep-sided hill that covers a relatively small area.

CALDERA. A volcanic crater much enlarged by explosive eruptions or collapse of the sides.

CANYON. Any vertical-walled valley cut by running water.

CHALK. A light, soft, white or gray limestone composed principally of shells of micro-organisms.

CINDER CONE. A cone, usually steep-sided, built up around a volcanic vent by accumulation of erupted materials.

CIRQUE. An amphitheaterlike hollow developed where a valley glacier forms.

CLASTIC SEDIMENTS. Sediments produced by rock destruction.

CLAY. Fine rock particles consisting mostly of silica, alumina, and water.

COASTAL PLAIN. A broad, low plain adjoining the sea; usually an emerged portion of the continental shelf.

COLUMNAR JOINTING. Jointing that produces long columns with a usually hexagonal or pentagonal cross section, as in basaltic rock.

CONE. A hill or mountain built up around a volcanic vent by erupted materials.

CONGLOMERATE. A sedimentary rock containing many stream-rounded pebbles and cobbles.

CONSEQUENT STREAM. A stream on a newly formed slope that takes a course determined by the dip of that slope.

CONTINENTAL SHELF. A gently sloping platform lying between some shorelines and the continental slope.

CONTINENTAL SLOPE. The steeply sloping surface between the continental shelf and the continental rise or ocean floor.

CONVECTION CURRENT. One of the slow currents believed to be produced in Earth's mantle by heat of the planet's interior.

GLOSSARY · 353

CORRASION. The erosion of bedrock by moving agents, such as running water, wind, or glacial ice.

CORROSION. Weathering or erosion due to chemical action.

COUNTRY ROCK. The regional rock into which igneous rock has been intruded.

CREEP. The slow, generally viscous, almost imperceptible downslope movement of soil or rock.

CROSSBEDDING. In sedimentary rock, any bedding or stratification that is inclined with respect to the main bedding planes.

CRUST. The outer shell of Earth, covering the mantle.

CRYSTALLINE ROCK. Rocks of crystalline structure, such as granite or schist, as distinguished from those of amorphous structure, such as sandstone or shale.

CUESTA. A ridge one side of which is much steeper than the other.

CUTOFF. The channel formed by a stream that cuts across the neck of an oxbow or horseshoe bend.

DEFLATION. The removal of material from a land surface by wind erosion.

DEGRADATION. The reduction of land by weathering, mass wasting, and erosion.

DELTA. A mass of sediments, often roughly triangular in form, deposited at the mouth of a river.

DESERT. Any terrain that receives less than 10 inches of rainfall annually.

DETRITUS. Material removed from rock masses by weathering and erosion; rock waste.

DIASTROPHISM. The processes by which Earth's crust is fractured and deformed.

DIFFERENTIAL WEATHERING AND EROSION. Weathering and erosion occurring at rates that vary because the rock resistance varies.

DIKE. A wall-like mass of intrusive igneous rock that cuts through the strata of sedimentary or metamorphic rock, or through a massive rock formation.

DIP. The inclination of any rock stratum or other flat surface with reference to the horizontal.

DOME. A rounded highland produced either directly by upfolding or by spheroidal weathering and erosion.

DRIFT. Glacial moraine.

DROWNED VALLEY. The portion of a valley that has been invaded by the sea as the result of a subsidence of the land or a rise of sea level.

DRY VALLEY. Any streamless valley, such as a wind gap or a valley cut by a stream that has been diverted (as in limestone country) to an underground course.

EROSION. The destruction of rock and removal of the debris by agents in motion —running water, waves or currents, windblown sand, or glacial ice.

ESCARPMENT. Any cliff or other steep slope that faces a level or gently sloping surface.

ESTUARY. The seaward end of a river valley, in which fresh and marine waters are mixed; usually, a drowned valley.

EXTRUSIVE ROCK. Any igneous rock formed by erupted materials on the surface of the crust.

FAN. An alluvial fan or a rock fan.

GLOSSARY · 354

FAULT. A fracture in bedrock that involves the movement of one rock mass with respect to another.

FAULT LINE. The intersection of a fault with the ground surface.

FAULT-LINE SCARP. An escarpment formed by differential erosion along a fault line.

FAULT SCARP. An escarpment formed directly by faulting.

FIORD. The seaward end of a valley that was much deepened by a valley glacier and later invaded by the sea.

FLOOD PLAIN. A terrain covered with sediments from periodic flooding by a stream.

FOLD. Any bend in rock structure caused by natural forces in the crust.

FRACTURE. Any break caused by stress or strain in bedrock.

FROST-WEDGING. The expansion of water within rock crevices as it freezes; also called frost-prying or ice-prying.

FUMAROLE. A hole or fissure in the ground through which volcanic gases escape.

GABBRO. An intrusive basic rock having relatively large crystals and consisting mainly of plagioclase feldspar and pyroxene.

GEOMORPHOLOGY. The science of landforms.

GEOSYNCLINE. A great trough of regional extent in which, over a long period of time, sediments may accumulate to depths of thousands of feet.

GEYSER. A hot spring from which, at intervals, columns of water and steam are erupted.

GLACIAL LAKE. Any lake that occupies a basin formed by the erosive or depositional action of a glacier.

GLACIER. Any large mass of ice formed by the compaction and recrystallization of accumulating snows; usually it spreads widely or moves downslope.

GNEISS. A strong metamorphic rock derived from granite or certain other igneous or sedimentary rocks.

GRABEN. A depression formed by the sinking of a crustal block, or the uplift of blocks adjacent to it, by faulting.

GRADATION. The leveling of a land surface by the processes of erosion, transportation, and deposition.

GRADIENT. The slope of any stream with respect to the horizontal.

GRANITE. A strong, light-colored intrusive igneous rock (occasionally metamorphic) composed mainly of potash, feldspar, and quartz.

GROUND WATER. Water underground, derived mostly from precipitation but to some extent from chemical reactions within the crust.

GUYOT. A flat-topped mountain rising from the ocean bottom.

HAMMADA. Bedrock exposed and polished by wind erosion in a desert.

HANGING VALLEY. A tributary valley that joins a main valley over an abrupt descent.

HOGBACK. Any sharp-backed ridge with very steep opposite slopes.

HOMOGENEOUS ROCK. Rock that is consistent in nature throughout.

HORST. A crustal block raised above adjacent blocks by faulting, or left standing high when adjacent blocks subside.

HOT SPRING. A spring whose waters are naturally heated to a temperature above that of the human body (about 98 degrees Fahrenheit).

HYDROLOGIC CYCLE. The set of processes by which water is evaporated from the surface of the ocean, precipitated onto the land, and returned to the ocean again by streams.

ICEBERG. A large block of ice that has become detached from the front of a glacier where it meets the sea.

ICE CAP. A mass of glacial ice that covers or nearly covers a group of highlands.

ICE SHEET. A large mass of glacial ice spreading out in many directions from a center of accumulation.

IGNEOUS ROCK. Rock formed by the cooling and solidification of molten materials.

INTERMITTENT STREAM. A stream that exists only part of the time, as during a rainy season.

INTRUSIVE ROCK. Igneous rock formed within the crust; also called plutonic rock.

ISOSTASY. The balanced condition in which large masses of relatively light rock tend to stand higher in the crust than masses of heavier rock.

JOINT. A fracture in bedrock caused by crustal movement or the cooling of an igneous mass after solidification.

KARST TOPOGRAPHY. A topography of sinkholes, blind valleys, and other solutional features developed on and within limestone.

LACCOLITH. A mass of intrusive igneous rock which has forced overlying rock upward to form a dome at or near the surface.

LANDSLIDE. A mass of earth plunging rapidly downslope.

LAVA. The term applied to magma after it has been erupted at the surface.

LIMESTONE. A sedimentary rock consisting mostly of calcium or magnesium carbonate.

MAGMA. Molten material within Earth's crust, derived from the mantle or from the melting of crustal rock.

MANTLE. The zone that lies immediately beneath Earth's crust, extending about halfway to Earth's center.

MASS WASTING. The downslope movement of surface rock masses by gravity.

MATURE STREAM. A stream at the stage marked by such features as a graded bed, a strong but deliberate flow, a moderately wide valley floor with gentle slopes, and a tendency toward balance between erosion of the trough and the transportation of sediments.

MEANDER. A loop in a stream, such as those commonly seen on flood plains.

MESA. A flat-topped erosional remnant of large size, usually with one or more steep sides.

METAMORPHIC ROCK. Any kind of rock that has been altered physically and chemically by heat, pressure, or infiltrating fluids.

MINERAL. Any of the naturally occurring chemical elements and compounds that make up the rocks of Earth's crust.

MONADNOCK. A hill or mountain isolated by erosion of a region to a peneplain.

MORAINE. Rock waste deposited by a glacier.

MOUNTAIN. A tract of land that at one stage of its development rises very high above its surroundings, having been produced by solid and fluid movements in the crust.

NORMAL FAULT. A fault in which the footwall appears to have been raised relative to the hanging wall.

OBSEQUENT STREAM. A stream that follows a course opposed to the dip of the rock strata or surface.

OUTCROP. Any appearance of bedrock at the ground surface; especially, the exposed edges of strata.

OVERTHRUST. A reverse fault with a very low dip and a displacement of great magnitude.

OXBOW. A horseshoe bend in a river's course.

PENEPLAIN. A terrain reduced almost to base level by erosion.

PLAIN. Any broad region of low slopes produced by deposition.

PLATEAU. Any horizontal or nearly horizontal strata genetically (structurally) related to nearby mountains.

PLAYA. A shallow desert basin in which runoff collects and is soon evaporated.

PLUTON. A mass of igneous rock formed within the crust.

PYROCLASTIC. Material blown out from a volcanic vent, consisting of fragmented lava and other rock.

QUARTZ. The mineral silicon dioxide (SiO_2), one of the commonest and most stable materials of the crust.

QUARTZITE. An extremely hard and resistant rock formed by the cementation of sandstone by silica, and frequently metamorphosed by shearing effects.

RAVINE. A steep-sided trench cut into a mountainside by running water, smaller than a valley but larger than a gully or gulch.

REJUVENATED STREAM. A stream which, after reaching maturity or old age in development, has been restored to youth by uplift.

RELIEF. The differing elevations of a land surface considered together.

RHYOLITE. An extrusive igneous rock similar to granite in its chemical composition but with finer crystals.

RIFT. A trench or trough that represents the intersection of a fault plane with the ground surface.

RIVER. Any large stream, especially a permanent stream larger than a brook.

ROCK. Any naturally formed mass of mineral matter, especially (but not necessarily) such material in consolidated form.

RUNOFF. Water that flows over the surface of the ground in streams or sheets.

SAND. Rock particles of any kind measuring from 2 to $1/16$ millimeters in diameter.

SANDSTONE. A sedimentary rock consisting mostly of sand grains cemented together with silica, lime, or iron oxide.

SCARP. Any cliff or steep slope of considerable linear extent.

SCHIST. Any of several kinds of metamorphic rock containing much mica and showing a wavy, uneven surface due to the orientation of minerals such as mica and hornblende.

SEDIMENT. Material deposited by water, wind, or ice.

SEDIMENTARY ROCK. Any rock formed naturally by the compaction and cementation of sediments.

SHALE. A sedimentary rock formed by the consolidation of clay and other fine-grained minerals.

SHEET EROSION. Erosion by waters flooding over the ground in broad sheets rather than separate streams.

SILICA. Silicon dioxide (SiO_2).
SILL. A sheet- or plate-like mass of igneous rock formed by magma intruding between strata of the country rock.
SILT. A clastic sediment consisting of rock particles from 0.005 to 0.05 millimeters in diameter.
SINKHOLE. Any depression caused by the sinking of earth; especially, one produced by solution in soluble rock such as limestone.
SLATE. A metamorphic rock, platy in appearance, formed from shale or fragmented igneous masses.
SLOPE WASH. Erosion of a slope by temporary sheets and streams of water; or, the material thus washed down.
SPRING. A flow of ground water emerging at the surface.
STILLSTAND. The condition of being stationary—that is, of being undisturbed by diastrophic activity.
STOCK. A steep-sided mass of plutonic rock covering less than 40 square miles.
STRATIFICATION. Layering or bedding.
STRATUM. Any distinct layer, or bed, in rock, especially sedimentary rock.
STREAM. A quantity of water or other material (such as ice or lava) flowing in linear rather than sheetlike fashion.
STRIKE. The direction of a horizontal line in a sloping plane.
STRIKE VALLEY. A valley that has been cut parallel to a tilted bed, fold, or fault.
SUBAERIAL EROSION. Erosion of a surface exposed to the direct action of the atmosphere—that is, on land.
SUBSEQUENT STREAM. A stream that has developed its course by headward erosion along a belt of weak rock.
SUPERPOSED STREAM. A stream which has held its course while cutting down through an underlying rock structure different from the structure in which the course was originally developed.
SWAMP. A saturated terrain covered with aquatic vegetation.
SYNCLINE. A downfold—that is, one in which the limbs dip inward toward the axis.
TALUS. Rock fragments from weathering and mass wasting that have fallen or slid to the bottom of a slope, forming there an outward-sloping accumulation.
THRUST FAULT. A reverse fault with a low angle of inclination to the horizontal.
TILL PLAIN. A plain built by accumulations of glacial drift.
UPLIFT. The elevation of a terrain with respect to sea level.
UPWARP. A low, broad anticline.
U-SHAPED VALLEY. A valley with a U-shaped cross section at the base resulting from glacial erosion.
VALLEY. A large trench cut into the land by a river; or, loosely, an elongated basin (rift valley) produced by faulting.
VALLEY GLACIER. A glacier confined to a valley (as distinguished from widespreading ice caps or ice sheets).
VOLCANIC ASH. Finely fragmented lava blown out from a volcanic vent.
VOLCANIC NECK. A body of lava which solidified in the pipe of a volcano when it became extinct, and then was left on the landscape as a tower or knob

after the weaker materials of the cone were eroded away.

VOLCANISM. The eruption of hot molten materials from Earth's interior through the crust.

VOLCANO. Any vent through which volcanic materials are erupted; or the hill or mountain formed by the accumulation of erupted materials around the vent.

WARP. A wide, shallow fold.

WATER CYCLE. *See* Hydrologic cycle.

WATER GAP. A transverse valley by which a stream passes through a mountain ridge.

WATER TABLE. The upper surface of the zone underground that is saturated with ground water.

WEATHERING. The mechanical and chemical processes that cause rock to disintegrate and decay when exposed to the weather.

WIND GAP. A water gap that has gone dry because of diversion of the stream that once occupied it.

YOUNG STREAM. Any stream of high gradient and rapid flow in a steep-walled valley with an ungraded bed, which is indicated by falls and rapids.

ZONE OF AERATION. The zone underground that is above the water table.

ZONE OF SATURATION. The zone underground that is below the water table.

Index

Asterisks (*) indicate pages containing illustrations related to the indexed subjects.

Aa, 163
Abajo Mountains, Utah, 154
Abert Lake, Ore., 272
Ablation, glacial, 224*-225
Abrasion platform, 310-311
Absaroka Range, Wyo., 86
Abyssal plains, 335-336
Acadia National Park, Me., 303*-304*
Acidic rocks, 37-41, 160
  see also Granite; Rhyolite
Aconcagua, 143
AdirondackMountains,N.Y.:
  basaltic intrusives, 35
  cirques, 236
  dome of anorthosite, 115, 151
  fault-line scarps, 128
  Giant Mountain, 151*
  lakes, 92, 250*, 255
  rectangular drainage, 92*
  slides, 57*, 151*

Agassiz, Lake, 194, 257
  terraces, 323
Agassiz, Louis, 10-11, 222
Akan National Park, Japan, 20*
Albert, Lake (Africa), 121
Aleutian Islands:
  Bogoslof, 172
  Kiska, 339*
  recent faulting, 125
  volcanoes of, 156*, 167, 172, 339*
Algae, coralline, 342-343
Alkaline lakes, 268-272
Allegheny Front, 140
Allegheny Plateau, 154
  see also Appalachian Plateau
Alluvial cones, 73, 77*
Alluvial fans, 193
  deserts and cold regions, 73

Alluvial fans—*cont.*:
  merging, 204
  Mohave Desert, 202*
  nature of, 203-204
  Rockies, 185
  Sierra Nevada, 146
Alluvial plains, 193-194
  Great Plains, 184*
  see also Deltas; Flood plains
Alpine-Himalayan system, 143-144
Alps:
  age of, 26
  Dolomites, 278
  extreme distortion in, 123, 144
  glaciers of, 248-249
  origin of, 144
  recent faulting, 125
  views of, 142*, 248*
Amargosa River, Calif., 83

359

INDEX · 360

Amazon River, 69
Andesite, 38-39, 40, 163
  nature of, 160
  origin of, 162
Andes Mountains, 143
  Pacific slope of, 331, 332
  recent faulting, 125
Angara River, Siberia, 271
Angeles National Forest, Calif., 88*
Anorthosite, 151
Antarctica, 201, 228-229
Antecedent streams, 86-87*
Anticlines, 107*, 108
  Rockies, 140
  valleys cut in, 83
Apache tears, 162
Apennine Mountains, Italy, 143
Appalachian Mountains:
  age of, 26
  Blue Ridge Province, 138-139
  depth of sediments in geosyncline, 133
  drainage, 113
  faulting, 113, 127
  Hall's investigations, 132
  history of, 136-140
  Newer Folded, 136-140
  old faulted terrains, 128
  overthrusts in, 127
  Piedmont Province, 136-137
  Ridge and Valley Province, 139
  topographic patterns, 113
  uplifts, 106
  West Virginia, 135*
  see also Appalachian Plateau; Blue Ridge; Folded Appalachians; Piedmont Province
Appalachian Plateau, 139-140, 187, 189*
  horizontal structure, 34
  recent uplift, 99
Arabian Desert:
  causes of aridity, 200
  dunes, 213*, 214, 215
  mountains, 218*
Archeozoic Era, 22, 24
Arches:
  desert, 207*
  sea, 307*, 308
  structural, 116
  see also Natural Bridges
Arêtes, 231, 237*
Aristotle, 8, 11
Arkansas River, 140
Artesian springs, 281, 295
Artesian wells, 253
Ascension Island, 337
Ash, volcanic, 162
  see also Lava dust
Atacama Desert, 200

Athabasca Glacier, Alberta, 228*
Atlantic City, N. J. (harbor), 315*
Atlantic Coastal Plain, 192
  dunes, 215
  recent changes, 26
Atlantic-Indian Ridge, 338
Atlas Mountains, 125, 143
Atolls. See Coral atolls
Ausable Chasm, N. Y., 74*-92
Australian Desert, 200
Auvergne province, France, 165, 174
Avalanches, 58-59
Azores, 172-173, 177, 337

Bacon, Sir Francis, 345
Badlands:
  causes of, 204
  plateau remnants, 190
  South Dakota, 205*
  stream spacing in, 91
Bahama Banks, 330
Bajada, 194
  Mohave Desert, 202*
  nature of, 204
Baikal, Lake (Siberia), 260
  in a graben, 121
  fresh water, 271
Baker, Mount, 167
Bandelier National Monument, N. Mex., 171*
Banff, Alberta, 77*, 123*, 234*
Bar, longshore, 310
Baraboo Range, Wisc., 43
Barchans, 211-215
  making of, 211*
  Saudi Arabia, 199*
  White Sands, N. Mex., 212*
Bar Harbor, Me., 98*
Barrier beach, (island) 97, 311*-313
Barrier reefs, 341-344
Barringer Crater. See Meteor Crater
Basalt, 35-39
  coasts of, 320
  columnar jointing, 36*, 46*, 164-165*, 181
  lava, 188
  plateaus, 175-177
  sea bottoms composed of, 328
  see also Basic rock
Base level, 16
Basement rocks, 346, 348
Basic rocks, 35-39, 159
Basin and Range Province, 144-146
  ages of scarps, 125-126
  causes of aridity, 200
  drainage in, 201

Basin and Range—cont.:
  dunes, 212*, 216
  lake plains, 194
  lakes, 260, 269 (map)
  recent faulting, 125
  streams affected by faulting, 102
Basins:
  blocked valleys, 261-262
  blowouts, 266
  clay-sealed, 282
  delta, 260-261
  developed in domes, 113-114, 117
  made by meteorites, 266*, 267
  made by plants and animals, 262-264*
  plunge, 260
  solution, 264*-265, 281-282
  structural, 257-260
  volcanic, 172, 265*, 267
  see also Grabens; Lakes; Sinkholes
Batholiths, 159*
  mountains carved from, 147-152
  nature of, 148
  Western (map), 149*
  see also Domes
Baymouth bar, 313
Bays, 308
Beach drift, 304
Beaches:
  Acadia National Park, 304*
  origin of, 309-310
  profile (diagram), 309*
  raised, 15*, 122, 246
  Virgin Islands, 298*
Beach ridges, 310
Beach scarp, 310
Bearfort Mountain, N. J., 232*
Bear Mountain, N. Y., 50*
Beartooth Mountains, Mont., 58*, 154, 187
Beavers, basins made by, 262
Benches:
  river-cut, 27*, 75, 83, 98, 99
  wave-cut. See Wave-cut benches
Bergschrund, 224*, 230
Berkshires, 129, 138
Berm, 310
Big-bang theory, 17
Big Hollow, Wyo., 210
Bighorn Mountains, 25, 141
Bighorn River, 140
Bights, 308
Bikini Atoll, 343
Black Forest Plateau, Germany, 121, 122, 187

INDEX · 361

Black Hills, 115
  dome, 150*
  drainage, 92
  granite forms, 48
  history of, 148-151
  relation to Rockies, 141
Blake Plateau, 331, 332
Blind valley, 281-282
Block mountains, 144-147
  see also Sierra Nevada;
    Teton Range; Wasatch
    Mts., etc.
Blowholes, 308
Blowouts, 209*-210, 266
Blue Mountains, Wash.,
  188, 189
Blue Ridge, 138*
  a catoctin, 155
  old faulted terrain, 128
  stream piracy, 89*, 90-91
  superposed streams, 86*
Blue Ridge Province, 138-
  139
Bogoslof Volcano, 172
Bolson, 194, 204
Bolson plain, 195*
Bonneville, Lake, 194, 260
  shorelines, 270*, 323
Bore, tidal, 306
Boston Mountains, Ark.,
  153*
Bournes, 281-282
Box canyons, 204
Boxwork, in caverns, 286*
Brahmaputra River, 86
Braiding, 76-77, 79*
Breakers, 301-304
Breccia, 38-39, 42
  fault, 119
  volcanic, 163
Brienz, Lake (Switzerland),
  262
Brittany coast, 320
Brocken, Germany, 155
Brook, mountain, 69-73
  see also Streams
Brooks Range, Alaska, 140
Bryce Canyon, Utah, 6*, 33,
  278
Buch, Leopold von, 10, 12
Bunch grass, 213
Bunker Hill, Boston, 241
Buttes temoines. See
  Pepinos

Calcium carbonate, 42, 276-
  277
Caldera:
  Crater Lake, Ore., 172,
    265*-267
  Krakatoa, 171
  nature of, 162
  Valles, 171
Calving, 229
Cambrian Period, 22, 24

Camel Rock, N. Mex., 51*
Canadian Rockies, 123*,
  140-141*
  see also Rocky Mountains
Canary Islands, 176
Canyons:
  box, 204
  origin of, 91, 204
  submarine, 331, 332-335
  see also Grand Caynon
Cape Cod:
  beach, 321*
  dunes, 214*
  erosion of, 316
  filling of lakes and
    harbors, 272
  glacial origin, 245, 321
  lakes, 256
  moraine topography, 241
  Provincetown spit, 312*
Cape Hatteras, N. C., 313,
  314*
Cape Henry, Va., dunes, 215
Capri coast, 320
Capulin Mountain, N. Mex.,
  28, 166
  view from, 184*
Carbonation of rock, 50-51,
  52
Carbon dioxide, 249, 276
Carlsbad Caverns, N. Mex.:
  age of, 287, 288
  Big Room, 284
  described, 294
  Papoose Room, 273*
Carolina Bays, N. C., 267
Caroline atolls, 343
Carpathian Mountains, 125,
  143, 144
Cascade Range, Calif., 166-
  167*, 172
  Columbia River ante-
    cedent on, 86-87*
Caspian Sea, 272
Castle Geyser, 181*
Castle Head, England, 174
Cataclysms, relief not
    formed by, 121
Catoctins, 155
Catskill Plateau, N. Y., 140,
  187
  cirques, 236
  mountains cut from, 154
  obsequent streams, 83
Caucasus Mountains, 144
Causse region, France, 279,
  290*, 295
Cave Mountain, W. Va.,
  135*
Cave of the Winds, Colo.,
  294
Caverns, limestone, 282-295
  age of, 288-289
  columns, 286
  deposited features, 284-
    289

Caverns—cont.:
  formation of, 282-284
  see also Caves
Caves:
  desert, 207
  sea, 303*
  see also Caverns
Cenotes, 294
Cenozoic Era, 22, 24
Central Asian Desert, 215
Central Fire, 12*
Chad, Lake (Africa), 203
Chalk:
  cliffs, 307*, 318*, 320
  permeability, 91
  pinnacles, 294*
  Selma, 44*
Challenger, H.M.S., 326
Challenger (Trieste) Deep,
  336
Chatham Strait, Alaska, 233
Chattahoochee River, 321
Chatter marks, 233
Chesapeake region estuaries,
  97, 320
Chevron folds, 108*
Chief Mountain, Mont.,
  125*-127
Chiltern Hills, England, 193
Chiricahua National Monu-
  ment, Ariz., 37*
Chubb Crater, Quebec, 267
Cimarron Valley, N. Mex.,
  40
Cincinnati Arch, 116, 291
Cinders, volcanic, 162
Cirques, 224*
  formation of, 230*-231,
    237*
  lakes in, 231, 254*
Clearwater Mountains,
  Idaho, 148
Climate:
  fluctuations of, 247-249,
    346
  influence on degradation,
    35-43, 51-52
  streams affected, 97
  see also Pleistocene Epoch
Clints, 278*
Cloud Peak, Wyo., 141
Coastal plains, 191-193*
  consequent streams on, 82
  cuestas of, 110
  European, 193
  lakes on, 259
  parallel drainage, 93
Coastline, defined, 310
Coast Range, 148*, 220*
Coast Ranges, Calif., 146
  region of recent faulting,
    125
Coasts, 296-323
  ages of, 26
  "Atlantic" type, 319-320
  basalt, 320

Coasts—*cont.*:
  diastrophic changes, 298
  "Pacific" type, 319-320
  raised, 15*, 133, 246
  sunken, 2*
  see also Shores
Coeur d'Alene Mountains, Idaho, 148
Col, formation of, 70
Colden, Mount, 57*, 183
Collapse sink, 278
Colorado Desert, 216
Colorado Plateau:
  cuestas, 110
  extent of, 187, 190
  intrusive rock masses, 35, 154, 182-183
  Powell's observations, 16
  rejuvenation of streams, 26, 97-98
  strath terraces, 99
  structural influence, 34, 106
  volcanic necks, 174, 175*
Colorado River:
  delta, 82
  deposition in Lake Mead, 272
  desert course of, 202
  in Grand Canyon, 98
  rejuvenation, 26, 97-98
  structural controls, 84-85
  volume of, 69
Columbia Plateau, 175-176*, 187*
  distinguished from Columbia Plateaus, 188
  scablands, 188-189
Columbia River, 187*
  antecedent on Cascades, 86, 87*
  plunge basin, 260
Columnar jointing, 36*, 46*, 164-165*, 181
Columns, limestone, 286
Como, Lake, 255
Composite scarp, 127
Cones. *See* Alluvial cones
Conglomerate, 38-39
  monadnock, 155*
  nature of, 41-42
  outcrop, 43*
  Shawangunk cuesta, 111*
Congo River, 69, 333*, 334
Connecticut River, 64*
  changing "age" of, 93-94
  superposition of, 86
Consequent streams, 112*, 191
Continental drift, 345
Continental shelves, 317*, 328-331
Continental shields, 134, 346
Continental slopes, 317*, 331-333

Continents:
  built up by volcanism, 17, 158, 345-348
  nature of, 133-134
  profiles compared to sea bottoms, 327*
Convection currents, 19, 336
Convection theory, 346-347
Cook Inlet delta, Alaska, 81*
Coral atolls:
  growth of, 340-344
  ponds on, 263
  Swain's Island, 324*
Coralline algae, 343
Cores of Earth, 18*
Cornwall Hills, England, 155, 320
Corrasion:
  by streams, 59-61, 70
  by waves and currents, 303
Corries. *See* Cirques
Cotswold Hills, England, 193
Coulees, 189
Coves, 308
Crater Lake, Ore., 172, 265*, 267
  see also Mazama, Mount
Craters of the Moon, Idaho, 161*, 166
  age of, 28
  pahoehoe, 163*
Crazy Mountains, Mont., 153, 166
Creep, 209
Crescentic fractures, 231-233
Crescentic gouges, 232-233
Cretaceous Period, 22, 24
Crevasses, 225
Crook Mountain, S. D., 153
Crossbedding, 42*
Crow Peak, S. D., 153
Crust, nature of, 18*-22
Cryptozoic Era, 22, 24
Crystal growth in weathering, 48
Cuestas, 111*, 186
  Appalachian, 113
  coastal plains, 193
  nature of, 110
Cumberland Plateau. *See* Appalachian Plateau
Currents:
  density, 305
  Georgia coast, 305*
  longshore (littoral), 305*, 306
  ocean, 304-305
  rip, 305*, 306
  shore, 304-306
  tidal, 305, 306
  turbidity, 329, 334-336
  wind-made, 305
Cuspate forelands, 313

Cusps, beach, 310
Cutoffs, 79
  lakes formed by, 260, 261*
  natural bridges formed by, 98
Cwm. *See* Cirques

Dams, natural, 261-262, 264
Danby Playa, Calif., 210
Danube River, 86
Darwin, Charles, 340, 343
Dating of rock:
  absolute, 23, 25
  by fossils, 22
Dead Sea, 16, 260
  horsts flanking, 122
  identified as in a graben, 15
  in Pleistocene, 271
  salt content of, 271-272
Death Valley:
  bolson plain, 195*
  desert pavement, 208*
  Devils Golf course, 203*
  faceted scarps, 124*
  graben, 121
  gullying, 59*
  ponds, 271
  streams, 83
  volcanic features, 167
Deccan Plateau, 176, 189
Deflation hollows, 209*, 210
Degradation, nature of, 32-33
  see also Erosion; Mass wasting; Weathering
Delaware River, 86
Delta plains, 194
Deltas:
  arcuate, 81*
  bird's-foot, 82
  conditions favoring, 321
  formation of, 81-82
  Gulf of Mexico, 329-330
  lakes on, 260-261
  Mississippi Delta, 68*, 95, 330
Depth of sounding, 326-327
Derwentwater, Lake (England), 257*
Deserts, 198-219
  angular landforms, 206
  blowouts, 209*, 210
  causes, 199-202
  caves in, 207
  Central Asian, 200
  defined, 199
  dunes and other sand accumulations, 210-216
  evolution of, 217-219
  "goblins," 217*
  lakes in pluvial periods, 260
  oases, 210
  pans, 210

INDEX · 363

Deserts—cont.:
  polar, 201
  pavement (serir), 208*
  rejuvenation, 219
  stream work in, 201*, 202-205
  subtropical, 200
  talus accumulations, 54, 145*, 145
  weathering, 205-206
  west-coast, 200
  windows and arches, 207
  wind work, 205, 206-219
Desert varnish, 208
Devil's Marbles, Australia, 47*
Devils Postpile, Calif., 36*, 165
Devils Tower, Wyo., 153, 165, 183*
Devonian Period, 22, 24
Diabase, 179, 181*
Diastrophism, 104-129
  base level and, 16, 105-107
  chronology of, 106-107
  coastal effects, 321-322
  erosion as influencing, 18
  evidences of, 106-107
  evolution of ideas about, 13-16
  general influence of, 105-107
  mountain-building, 132-136
  see also Faults; Subsidence; Uplift
Diego Alvarez Island, 337
Dikes, 159*
  Arabian Desert, 217*
  nature of, 181-182*
  streams in, 73*, 183
  volcanic cones, 174-175
Dinaric Mountains, 143
Dinosaur National Monument, 9*
Diorite, 38-39
Dip and strike, 110*
Dismal Swamp, N. C., 263
Distributaries, 82
Doline, 278
Dolomite, 38-39
  nature of, 42
  solubility of, 277
  see also Limestone
Dolomitic Alps, Italy, 143*, 278
  coral reefs in, 344
  influence of jointing, 34
Domes, 113-118
  batholithic, 114-115, 147-152
  drainage on, 92-93, 116-117
  salt, 117*-118
  volcanic, 168*-169

Dover Cliffs, England, 116, 318*
Downcutting by streams, 69-73
Downs, chalk (England), 277
Drainage:
  desert, 201, 217-219
  North American, 66*
  patterns, 91-93
  systems, 87-93
  volcanoes, 165-166
Dreikanter, 206-207
Drift. See Glacial drift; Glacial moraine
Dripping Springs Escarpment, Ky., 291
Dripstone, defined, 284
Drumlins, 240-241, 245, 246*
  Boston Harbor, 321
Drummond, Lake, 263
Dry Tortugas, 343
Dry valley, 281-282
Dune ridges, 310
Dunes, 210-216
  barchans, 199*, 211-215
  European areas, 215
  parabolic, 212*
  seifs, 213*-214
  underwater "dunes," 330
  whalebacks, 214
Dust, volcanic, 162, 170
Dust Bowl, 209
Dust storms, 209
Dutton, Clarence, 132

Earth:
  interior, 18-22
  origin, 17
  unique conditions on, 31
Earthflows, 57
Earthquake Lake, Mont., 97, 263*
Earthquakes:
  Alaskan (1964), 119-120
  cause of, 119
  rockfalls caused by, 55*
  San Francisco, 121
  submarine, 336, 337, 338
  West Yellowstone, 55*, 97, 263*
Earthquake waves, study of, 18
East Pacific Rise, 338
Echo-sounding, 327
Edinburgh Castle, Scotland, 174
Elbert, Mount (Colo.) 141
El Capitan, Tex., 344*
Elephant backs, 205*
Elkhorn Peak, S. D., 153
Emerald Bay, Calif, 319*
Emmons Glacier, Wash., 247*
Eniwetok Atoll, 343

Eocene Epoch, 24
Erg, 208-209, 211
Erie, Lake, 323
Erosion:
  base level and, 16
  cycles, 102-103
  defined, 33
  diastrophism influenced by, 19
  evolution of ideas about, 8-11
  most landforms shaped by, 106-107
  on various rock structures, 34*
  rates of, 3
  submarine, 333-336
  summary of, 59-61
  see also Glaciers; Rivers; Solution; Streams; Wind; etc.
Erratics, 222, 242, 243*
Esker, 239-240*, 245, 246*
Estuaries, 321
  Chesapeake region, 97, 320
  Maine Coast, 98*
  origin of, 97
Etna, Mount, 178
Everglades, Fla., 191*, 279*
Exfoliation, 46*-48
Extrusive rocks, 35, 38-39

Faceted pebbles, 206-207
Faceted spurs, 231, 236
Fall line, 84
Fans. See Alluvial fans
Fault-block mountains. See Block mountains
Fault breccia, 119
Fault gouge, 119
Fault-line scarps:
  nature of, 127, 128*, 129
  Ramapo, 127-128, 129*
Faults:
  abyssal plains, 336
  continental slopes, 332
  defined, 119
  karst topography, 277
  landscapes on, 121-129
  limits of, 119-121
  patterns of, 119*, 121
  recent, 125-127
  San Andreas Fault, 118*, 121
  Sandia Mountains, N. Mex., 120*
  stream in a fault, 101*
  streams affected by, 97, 102, 118*
Fault scarps:
  distinguished from fault-line scarps, 127-128*
  faceting of, 124*
  nature of, 121
  old and young, 145-146

INDEX · 364

Fayal, Azores, 172-173
Fedtchenko Glacier, India, 248
Felsenmeer, 48*-49
Fiji atolls, 343
Fingal's Cave, Scotland, 165, 320
Finger lakes, 256
Finlayson Channel, B. C., 233
Fiords, 233, 318
  Milford Sound, 236*
  regions of, 320
  Tracy Arm, 148*
Firn, 223
Fissure flows, 175-177
  Columbia Plateaus, 175, 176*, 187*
  Deccan, 176, 189
  Iceland, 176, 190
  Ireland, 190
Flamborough, England, 316
Flatirons, 142
Flattop Plateau, Colo., 187, 190
Flood plains, 193
  lakes on, 260
  Mississippi, 94-95
  origin of, 79-80
  White River, 80*
  Yukon Flats, 96*, 261*
Flowstone, 285, 287
Flume, the, 73*, 183
Folded Appalachians, 139
  erosion on, 111-113
  structural plains, 190
  structure, 136-137
  subsequent streams, 84
  superposed streams, 85*-86*
Folded mountains, 136-144
  see also Folded Appalachians; Rocky Mountains; etc.
Folds:
  chevron, 108*
  landscapes on, 34*, 110-118
  nature of, 107*-109*
  syncline, 14*
Fontaine de Vaucluse, France, 281
Foraminifera, 342
Foreshore, 310
Fossils:
  ages of rocks indicated by, 22
  early ideas about, 9
Fractures:
  crescentic, 231-233
  lunate, 233
  nature of, 119
  see also Faults; Joints
Franconia Notch, N. H., 30*, 73*
Fringing reef, 342

Front Range, Colo., 139*, 142
Frost flowers, 285*, 287
Frost polygons. See Patterned ground
Frost prying, 48, 60*
Fujiyama, 11, 160*, 161
Fumaroles, 161-162, 177-178
Fundy, Bay of, 306
Furnace Creek, Calif., 83

Gabbro, 38-39
Galilee, Sea of, 16
Gaping Ghyll, Yorkshire, 295
Garden of the Gods, Colo., 104*, 142
Garret Peak, Wyo., 141
Gassi, 214
Genesee River, N. Y., 99*
Geneva, Lake (Switzerland) 82, 256
Geologic time, 26-29
  divisions, 22-24
  Hutton's view, 14
  table, 24
George, Lake (N. Y.), 128, 260
Georges Bank, 329
Geosyncline, 132-133
  Alpine-Himalayan, 144
  Appalachian, 136
  Rocky Mountain, 140
Geysers, 178*-179, 181*
Geysers, The (Calif.), 178
Giant's Causeway, N. Ireland, 165*, 320
Gibraltar, tombolo at, 313
Glacial drift, 238-245
  ocean bottom accumulations, 335, 329
  shore deposits, 321
  see also Glacial moraine
Glacial moraine, 222, 239-245, 246*
  bands on glacier, 220*, 225
  see also Till, glacial
Glacial stairs, 233
Glacial troughs, 231, 233, 234*-235, 237*
Glaciated plains, 195-196, 197*
Glaciation:
  Agassiz as student of, 10-11
  alpine, 234, 248*
  continental shelves, 329
  evidences recognized, 222, 230
  ice sheet, 244*
  streams deranged by, 97
  see also Glaciers
Glaciers, 220-249
  deposits, 238-245
  erosion (diagrams), 237*

Glaciers—cont.:
  fluctuations, 247-249
  grooving, 239*
  land depressed, 246, 247, 320, 329
  modern, 222
  moraine bands, 220*, 225
  movement of ice, 224
  North Sawyer, 10*
  outlet, 229
  piedmont, 226-229
  polar, 228-229
  rock, 55
  valley type, 220*, 222-226
  valley forms, 231, 237*
  volcanoes, 166
  see also Glaciation and other related entries
Glen Canyon Dam, 272
Gneiss, 38-39, 43-44
  banded, 45*
Gobi Desert, 144, 200
"Goblins," 32*, 51*, 217*
Gouges, crescentic, 232*-233
Grabens:
  Gulf of California, 322
  lakes in, 259*-260
  nature of, 121, 128*
  Persian Gulf, 322
  Red Sea, 322
  Sea of Galilee, 16*
  streams in, 83
  submarine, 332
  Thingvellir, 337*
Grampian Mountains, Scotland, 23*
Grand Banks, 329
Grand Canyon:
  age, 26, 28
  benches, 27*, 99
  cut by a rejuvenated stream, 98
  future of, 85
  Ribbon Falls, 101*
  structural controls, 84-85
Grand Coulee, 260
Grand Tetons. See Teton Range
Granite, 38-39
  described, 37
  exfoliation, 46*
  jointing in, 46*
  landforms, 37-38, 48
Gravestones, weathering of, 52
Gravity, 31-33
  see also Mass wasting
Great Basin. See Basin and Range Province
Great Dismal Swamp, 263
Great Glen, Scotland, 256*
Great Lakes:
  development (maps), 258*
  raised beaches, 246

INDEX · 365

Great Lakes—cont.:
  salinity, 272
  wave-cut terraces, 323
Great Plains:
  alluvial fans, 193
  blowouts, 210
  drainage, 93
  dune areas, 216
  horizontal structure, 34, 106, 185
  Panhandle, 192*
  Scotts Bluff, Nebr., 4*
  temporary lakes, 266
Great Rift Valley, Africa, 121, 152*
Great Salt Lake, Utah, 270* 272
  see also Bonneville, Lake
Great Sand Dunes National Monument, 215, 216*
Great Sandy Desert, Ore., 218
Great Smoky Mountains, 138, 152*
  gneisses, 44
  old faulted terrain, 128
Great Valley, Calif., 146
Great Valley of Appalachian region, 42, 291, 293
Great Whin Sill, 181
Greek island horsts, 122*
Greenland ice cap, 201, 229
Green Mountains, 138
  old faulted terrain, 129
  valleys in marble, 44
Green River, Ky., 291
Grinnell, Lake, 255*
Grinnell Glacier, 255*
Groins, 314, 315*
Gros Ventre Mountains, Wyo., 97, 262
Ground moraine, 240
Ground water, 69-70, 276-277
  amount in crust, 69-70
  hot springs and geysers, 178-179
  lakes, 252-253
Guadalupe Mountains, Tex., 294, 344*
Guettard, Jean, 9
Gulf Coast:
  erosion on, 320-321
  salt domes, 118
Gulf Coastal Plain, 192-193
Gulf of California, 322
Gulf Stream, 329
Gullies, 70-73, 91, 204
  Death Valley, 59*
  on fault scarp, 124
  systems, 88
Gull Pond, Cape Cod, 258*
Gunnison River, Colo., 57, 262
Guyots. See Seamounts
Gypsum rosettes, 288*

Hackensack Meadows, N. J., 174
Halemaumau fire pit, 169
Half Dome, Yosemite, 48
Hall, James, 132
Hammada, 208-209
Hanging valleys:
  glacial, 75*, 231
  on fault scarps, 124
Harbor engineering, 313-316
Harney Peak, S. D., 141
Harney Playa, Ore., 209*, 268*
Harpers Ferry, W. Va., 78*
Hartland Point, Devonshire, 296*
Harz Mountains, Germany, 103-118
  see also Black Forest Plateau
Hawaiian Islands:
  age, 28
  erosion, 174
  Koko Head, 340*
  shores, 322*
  volcanoes, 168*-169, 338
Haystack hills. See Pepinos
Helictites, 284, 285*, 286
Henry Mountains, Utah, 153
Herculaneum, 166, 170
Herodotus, 8, 326
Higgins Pond, 258*
Highwood Mountains, Mont., 154
Himalaya Mountains:
  age of, 26, 28
  origin of, 144
  recent faulting, 125
  still rising, 144
Hogbacks, 186
  defined, 113
  Garden of the Gods, 104*, 142
  Rockies, 141-142
  Sinclair dome, 116*
Hogsty Reef, Bahamas, 343
Holyoke Range, Mass., 86, 94, 167
Homocline, 108, 113
Hood, Mount, 167
Hoosac Range, Conn., 138
Hoover Dam, 272
Horn peaks, 237*
  Alps, 248*
  formation, 231
  Matterhorn, 142*, 144
  Teton Range, 235
Horsts, 121-122*, 144-145
Hot springs, 177*, 178
Hot Springs, Ark., 178
Hot Springs, Ga., 166
Hourglass valleys, 124*
Huascaran, Mount (Peru), 59
Hubbard Glacier, Alaska, 248

Hudson Canyon, 334*, 335
Hudson Highlands, 94*, 128-129*
Hudson Palisades, 179, 181*
Hudson River, 82, 86, 94*, 233
Hudson Valley (a fiord), 233
Humboldt River, Nev., 202
Humidity. See Climate
Hums. See Pepinos
Hutton, James, 67, 132
  erosion as viewed by, 9-10
  geologic time as understood by, 14
Hydration of rock, 50, 52
Hydrologic cycle, 32, 60-61, 69

Ice age. See Pleistocene Epoch
Ice arch, 226, 228*
Icebergs, 229
Ice caps, 228-229
  see also Ice sheets
Icefall, 224*
Iceland:
  fissure flows, 176, 190
  rift zone, 337*
Ice-marginal lakes, 257
Ice pillars, 225
Ice prying, 48, 60*
Ice-scoured plains, 195
Ice sheets, 223*, 235-236, 244*
  see also Ice caps
Ice shelf, 229*
Idaho Batholith, 148
Igneous rocks, 19, 35-41, 159-165
  see also Basalt; Granite; etc.
Ignimbrite Plateau, N. Z., 189
Indiana Dunes, 216
Indian Creek, Ind., 90
Indus River, India, 86, 87
Ingleborough district, England, 289, 295
Ingraham Glacier, Wash., 227*
Inlets, 308
Inselbergs, 155, 219
Insolation, 249
Interlaken, Switzerland, 262
Intrusive rocks, 35, 38-39, 159, 179-183
  see also Batholith; Laccolith; Stock, etc.
Inverted topography:
  domes, 117
  faults, 127, 128*
  folds, 111-113
Island arcs, 125, 173-175,

346
Island Beach, N. J., 311*
Islands:
  oceanic, 338-344
  sand barriers, 311*-313, 320
  sea stacks, 308, 319*
  volcanic, 37, 172-175
Isostasy, 132, 328

Jackson Hole, Wyo., 147
James River, Va., 91
Jasper National Park, 228*
Jemez Volcano, N. Mex., 171
Jetties, 314, 315*
Joints, 33-34
  columnar, 36*-39, 164-165*, 181
  effect on talus, 54*
  granite, 37, 46*
  pinnacles due to, 6*, 278
Jordan Valley, 16*
Josephine, Lake (Mont.), 255*
Judith Mountains, Mont., 153
Jungfrau, Alps, 248*
Jurassic Period, 22, 24
Juvenile water, 178

Kaaterskill Creek, N. Y., 83, 89
Kaietur Falls, Venezuela, 84
Kalahari Desert, 200
Kalambo Falls, Africa, 121
Kame, 240, 241, 242*, 245*, 246*
Kame terrace, 240
Kar. See Cirques
Karren, 278*
Karst, the, 275-276
Karst cycle, 289-290, 292*
Karst regions, 290-295
Karst topography, 274*-295
Karst valley, 282
Karst windows, 281
Katahdin, Mount, 151
Katmai, Mount (Alaska), 167, 178
Kettle lakes, 256, 258*
Kettles, 241, 246*
Kīlauea Volcano, 11, 169
Kilimanjaro, Mount (Africa), 248
King River, Calif., 146, 261
Kiska, Mount (Alaska), 339*
Kivu, Lake (Africa), 262
Klamath Lakes, 126
Klamath Mountains, Calif., 146
Klippes, 144
  Chief Mountain, 125*-127
  nature of, 119, 126-127

Koko Head, Hawaii, 340*
Krakatoa Volcano, 170-171, 299
Kwajalein Atoll, 343

Laacher See (Platten See), Germany, 267
Laccoliths, 153-154*, 159*, 183*
  Colorado Plateau, 114, 116*
Lacustrine plains, 194, 271*
Lagoon, 313, 343
Lahontan, Lake, 260
Lake District, England, 255, 257*
  batholithic dome, 115
  superposed streams, 86
  volcanic neck, 174
Lake plain, 194, 271*
Lakes, 250-273
  blocked-valley, 261-262
  cirque, 254*, 255
  coastal-plain, 259
  coral-island, 263
  currents in, 323
  defined, 252
  delta, 260-261
  desert, 194, 204, 268*, 271, 272
  extinction of, 272-273
  glacial, 251*, 254-257
  graben, 259*-260
  ice-marginal, 257
  in basins made by plants and animals, 262-264
  in structural basins, 257-260
  in wind-made basins, 265-266
  kettle, 256, 258*
  moraine, 256
  oxbow, 80*, 96*, 260, 261*
  piedmont, 255-256
  rock-basin, 251*
  shore erosion in, 323
  sinkhole, 264*-265
  tides in, 323
  trough, 255-256*
  tundra, 263
  types of basins, 252
  waves on, 323
Land's End, Colorado Plateau, 32*
Landslides, 58*, 97, 263*
Lapiés, 278*
Lapilli, 162
Laramie Range, Colo., 141
Lascaux Caves, 295
Lassen, Mount (Calif.), 166
Lassen Volcanic National Park, 97, 262
Lateral moraine, 239
Laurentian Mountains, Quebec, 25, 26, 129

Lava, 162-165
  acidic, 162, 163
  basic, 162, 188
  caves and tunnels, 163, 164*
  changes in kinds erupted, 162
  dust, 40*, 163, 171*
  kinds of, 159-161
  magma distinguished from, 159
  plains, 177, 187
  plateaus, 187, 188
  see also Igneous rocks; Volcanoes
Lava Beds National Monument, 164*
Lehman Caves, Nev., 273*
Le Puy, France, 13*
Levee, natural, 80, 336
"Lily Pads" (travertine), 287
Limestone:
  chalk, 44*, 116, 277, 294*, 307*, 318*, 320
  coasts, 320
  land-forming characteristics, 42-43, 274*-295
  occurrence, 290
  permeability, 91
  pinnacles, 278, 294*
  solution process, 276-277*
  surface sculptures, 278
  see also Caverns, limestone
Limestone country, 274*-295
Linville Falls, Va., 72*
Lisbon (Portugal) tidal wave, 299
Little Sun Dance (Green Mountain), Wyo., 153
Littoral currents, 305*, 306
Lituya Bay, Alaska, 299-300
Loch Ness, Scotland, 255, 256*
Loch Oich, Scotland, 256*
Loess, 210, 244*, 245
Long Island, N. Y. (glacial origin), 241, 321
Longshore bar, 310
Longshore currents, 305*, 306
Longs Peak, Colo., 139*
Lopolith, 181
Lost River Cavern, Ind., 288
Lost rivers, 89-90, 279-281, 288
Louise, Lake (Alberta), region of, 60*
Low-tide terrace, 310
Lucerne, Lake (Switzerland), 259
Lugano, Lake (Italy), 255
Lunate fractures, 233
Lyell, Charles, 10, 12, 15

INDEX · 367

MacDonald, Lake (Mont.), 255
MacDonnel Ranges, Australia, 114*
Mackinac Island, Mich., 323*
Madison River, Mont., 263*
Maggiore, Lake (Italy), 255, 259
Magma:
  nature of, 19, 159, 162
  reservoir in Vesuvius, 170
Malaspina Glacier, Alaska, 248
Mammoth Cave:
  age of, 287, 288
  description, 291-293, 294
  gypsum rosettes, 288*
  rate of solution, 288
Mammoth Hot Springs, 177*, 178
Manly, Lake (Calif.), 203
Mantle, 18*-19, 132
Marble, 38-39, 43-44, 52
Marcy, Mount (N. Y.), 151
Mariana Trench, 336
Marine-cut terrace, 310-311
Marshall atolls, 343
Martinique, W. I., 172
Massanutten Mountain, Va., 90*
Mass wasting:
  defined, 33
  in valley widening, 71
  nature of, 52-56
  on mountains, 135-136
Mato Tepee. See Devils Tower
Matterhorn, 142*, 144, 231
Mauna Loa, 158, 168*-169, 188, 338
Mazama, Mount (Ore.), 172, 265*
  age of, 28
  tuff from, 163
McKinley, Mt., 1*
Mead, Lake (Ariz.), 82, 272
Meanders:
  Colorado Plateau, 97-98
  origin of, 79-80
  White River, Ark., 80*
Meander scars, 261*
Mears Glacier, Alaska, 226*
Medicine Bow Range, Colo., 141
Mesa de Maya, 17
Mesas, 111, 155, 206
Mesozoic Era, 22, 24
Messier Channel, Chile, 233
Metamorphic rocks, 38-39, 43-44
Meteor Crater, Ariz., 266
Meteorites, 335
Meuse River, France, 86, 97
Mfumbiro Mountains, 262

Michigan, Lake, 256
Mid-Atlantic Ridge:
  expansion of crust, 346-347
  volcanic activity, 172, 176
Midland Valley, Scotland, 121
Mid-Ocean Ridge, 336-338
  map, 326
Midway Island, 169, 343
Milford Sound, N. Z., 236*
Miocene Epoch, 24
Misfit streams, 97, 100*
Mississippian Period, 24
Mississippi Delta, 55
  bird's-foot nature of, 82
  Dennis Pass, 68*
  depth of, 330
Mississippi River:
  changing "age" of, 94-95
  deposition on Gulf Plain, 193
  Pleistocene diversion of, 244
  work of, 67-68
Mississippi Valley:
  age of, 68
  peneplains, 102
Missouri River, 244
Mogotes. See Pepinos
Mohave Desert, 202*, 210, 216
Mohole project, 348
Mohorovicic discontinuity, 336, 348
Molokai, H. I., 322
Monadnock, Mount (N. H.), 155
Monadnocks, 154-155*
  limestone, 290, 293*
Monocline, 107*, 108, 109*
Mono Craters, 146
Mono Lake, Calif., 272
Montague Island, Alaska:
  breakers on shore, 302*
  fault, 119-120, 126*
Monterey Canyon, Calif., 334
Moon, origin of, 348
Moraine. See Glacial moraine
Moraine lakes, 256
Moro, Abbe (De crostacei), 12
Morro Bay, Calif., 312*
Moulin, 240
Mountains, 130-156
  batholithic, 147-152
  birth of, 132-136
  block, 144-147
  chains of, 134
  chronology of, 134-135
  cores of Appalachians, 128-129
  desert, 125-126, 206, 218*

Mountains—cont.:
  erosion as a cause of ruggedness, 135-136
  folded, 136-144
  of circumerosion, 154-155*
  old attitudes toward, 131-132
  perennial nature of, 134
  related to continents, 133-134
  stocks and laccoliths, 153-154*
  submarine, 336-338
  types compared, 135
  volcanic. See Volcanoes
Mud flats, 313
Mudflows, 57
  desert, 204-205, 206
  Slumgullion, 262
  volcanic, 165-166, 170
Mudpots, 178
Muir Glacier, 249
Mushroom rocks. See Pedestal rocks
Myozinsyo Volcano, 173
Mythen Peak, 144

Nashville Dome and Basin, 83, 117
Natural bridges, 98, 290
Natural Chimneys, Va., 278
Natural levees, 80, 336
Navajo Desert, Ariz., 214
Navajo Mountain, Utah, 114, 154*
New England Upland. See Blue Ridge Province
Newer Folded Appalachians. See Folded Appalachians
New Market Gap, Va., 90*
Niagara Falls, N. Y., 84*
Nickpoints, 99
Nihau, H. I, 322
Nile Delta, 8, 82, 210
Nile River, 202
  see also Nile Delta
Nivation, 223, 236
Noah's flood, 8, 13-14
Normandy coast, France, 307*
North Cape, Norway, 245*
North Downs, England, 116
North Platte River, 77
North Sawyer Glacier, Alaska, 10*
Notch, wave-cut, 303*
Nuée ardente, 172
Nunataks, 229, 246*
Nyasa, Lake, 121, 260

Oases, 210
Obsequent streams, 83*, 112*
Obsidian, 38-39, 163

INDEX · 368

Ocean bottoms. *See* Sea bottoms
Ocean Island, 169
Oceans, origin of, 17
Ocean waves. *See* Waves
Offshore zones, 310
Ohio River, 244, 279
Okanogan Range, Wash., 230*
Okeechobee, Lake (Fla.), 259
Old Man of the Mountain, 28*
Olga, Mount (Australia), 155*
Oligocene Epoch, 24
Olympic National Park, 308*
Ordovician Period, 24
Organic reefs, 340-344
Ouse River, England, 91
Outliers, 2*, 186
Outwash plain, 240, 245
Overthrusts, 119, 126-127
Owens Lake, Calif., 272
Owens River, Calif., 202
Oxbow lake, 80, 260, 261*
Oxbows, 79-80
Oxidation of rock, 49, 52
Ozark Dome, 151
Ozark Mountains, 153*

Pacific Coast, recent changes in, 26
Pacific Rise, 338
Pack ice, 229*
Pahoehoe, 163*, 164
Painted Desert, Ariz., 30*, 40*, 190
Palau Islands, 341*
Paleocene Epoch, 24
Paleozoic Era, 22, 24
Palouse Falls, Wash., 176*
Panamint Mountains, Calif., 195*
Panhandle, Tex., 192*
Pans, 210
"Parachute," 273*
Paradise Range, Nev., 145*
Paraná Plateau, 176, 189
Paris Basin, 93, 193
Park Range, Colo., 141
Patch reefs, 343
Paternoster lakes, 254-255
Patterned ground, 56*
Patterns, drainage, 91-93
Pedestal rocks, 30*, 51*, 217*
Pediment, rock, 204
Pelé, the fire goddess, 11, 169
Pelée, Mount (Martinique), 172
Pelé's hair, 163
Pelé's tears, 162
Peneplains, 102-103*

Peninsular Ranges, Calif., 146
Pennine Hills, England, 91, 144, 278*
Pennsylvanian Period, 24
Pennyroyal Plateau, Ky., 291
Pepinos, 290, 293*, 295
Peridotite, 38-39
Permian Period, 24
Persian Gulf, 322
Petroleum, 342
Phacoliths, 181
Piedmont lakes, 233, 255-256
Piedmont Province, 136-137, 192
  peneplain, 102, 103*
  uplands, 84, 128
Pikes Peak, Colo., 49, 155
Pillow lava, 163, 174
Pindus Mountains, Greece, 143
Piracy by streams, 88-91
Pit craters, 161, 168
Placid, Lake (N. Y.), 92, 250*
Plains:
  abyssal, 335-336
  alluvial, 184*,-186, 191-196
  glaciated, 195-196, 197*
  lacustrine, 271*
  lava, 177, 187
  outwash, 240, 245
  rock-floored, 195, 208-209
  stripped (structural), 190
Plants, rock desetruction by, 49*
Plateaus, 184-190, 347*
  basalt, 175-177, 187*, 189
  distinguished from plains, 191
  erosion on, 9*, 32*, 186*
  structural, 190
  submarine, 331
  see also Appalachian Plateau; Colorado Plateau; *etc.*
Plattekill Creek, N. Y., 83
Playa, 194, 268*
Playa lakes, 204, 271
Playfair, John, 87
Pleistocene Epoch, 24
  desert landforms influenced, 205, 269-270
  events of, 222
  extent of glaciation, 221, 223*
  pluvial period, 205, 260, 269-270
Pliny the Elder, 12
Pliocene Epoch, 24
Plug dome (spine), 163, 172
Plunge basin, 231, 260

Plutonic rock. *See* Intrusive rocks
Plutons, 159, 179-183
Pluvial period, 205, 260, 269-270
Pocket valleys, 281-282
Pocono Mountains, Pa., 140, 154
Po Delta, 82
Polar deserts, 201
Poljes, 279
Pompeii, burial of, 163, 170
Ponores, 279
Pontchartrain, Lake (La.), 261
Potholes, 71*, 95*
  drilling of, 70
  glacial, 236
  in caverns, 284
Potomac River, 90
  junction with Shenandoah, 78*
  water gap in Blue Ridge, 86*
Po Valley fans, 193
Powell, John Wesley, 10, 16
Pressure ridges, 225
Profile rock, 28*
Providence Canyons, Ga., 294*
Provincetown spit, 212*
Puerto Rico Trench, 336
Pumice, 38-39, 41, 163
Punkaharju region, Finland, 240*
Puy du Dome area, 28
Pyrenees Mountains, 143
Pyroclastics, 160, 162

Qattara Depression, Egypt, 210
Quartz, 38-39, 50, 206
Quartzite, 38-39, 43-44

Radioactive dating, 25
Rainfall during geologic time, 60-61
Rainier, Mount, 166
  glaciers, 229*, 247*, 248-249
  radial drainage, 92
Rain shadow, 200
Ramapo Mountains, 44, 128-129*, 138
Ramp valleys, 121
Rapids, 72, 99
Raton Mesa, 177
Ravines. *See* Gullies; Valleys
Rebound, postglacial, 246, 247, 320, 329
"Red rains," 209-210
Red River timber-raft dam, 264
Red Rocks, Colo., 142
Red Sea, 260, 322
Reefs, organic, 340-344

INDEX · 369

Refraction, wave, 303-304
Rejuvenated streams, 97-98, 99*, 219
Relief:
  extremes of, 22-23
  influenced by elevation, 33, 35
Resequent streams, 83*, 112*
Residual mountains, 154-155*
Resistance, rock, 33-35, 71-72, 320
Rhine Delta, 82, 261
Rhine Valley, 83, 121
Rhône Delta, 82
Rhône Glacier, 249
Rhyolite, 37*, 38-39
  described, 40, 163
  flows, 188
Ribbon Falls, Grand Canyon, 101*
Ridge and Valley Province. See Folded Appalachians
Ries Crater, Bavaria, 267
Rifts, 121, 122
Rill marks, 310
Ring dikes, 174-175
Rio Grande, 202
Rio Grande Ridge, 338
Rip channels, 310
Rip currents, 305*, 306
Rip head, 306
Rip tides. See Rip currents
Rises:
  ocean, 338
  springs, 281
Rivers, 65-103
  ages of, 93-96
  antecedent, 87*
  benches cut by, 27*, 75, 83, 98, 99
  blocked, 97, 261-262
  deranged, 97, 102, 118*, 122, 242, 244
  grabens occupied, 83
  growth of, 70
  interruptions of development, 96-102
  "lost," 89-90, 279-281
  mature, 94*
  misfit, 97, 100*
  old, 96*
  old ideas about, 66
  superposed, 85*-86*
  swinging, 78
  total water in, 69
  valleys, 73-78
  young, 95*
  see also Streams
Roanoke River, Va., 91
Robson, Mount, 140
Roches moutonnées, 236, 237*, 246*
Rock creep (drift), 53*-56
Rockfalls, 55*
  see also Mass wasting

Rock fan, 204
Rock-floored plain, 195, 208-209
Rock "flour," 226, 254
Rock glacier, 55
Rock resistance, 33-35, 71-72, 320
  see also Erosion; Weathering
Rocks:
  as land-formers, 35-46, 91
  basement, 346, 348
  classification chart, 38-39
  distortion of, 107-110
  oldest, 25
  relative ages indicated by fossils, 24
  see also other Rock entries
Rock steps, glacial, 233
Rock structure:
  coastal effects, 318-319
  horizontal, 32*, 33-35, 185
  solution influenced by, 277
  streamwork guided by, 71-72, 82-85
Rocky Mountains:
  age of, 26
  folded, 140-143
  Front Range, 139*
  glacial sculptures, 235
  history of, 106, 140-143
Rosettes, gypsum, 288*
Ross Ice Shelf, Antarctica, 229*
Round (Palisade) Butte, Mont., 154
Rundle, Mount (Alberta), 123*
Ruwenzori horst, 144-145

Saddle, formation of, 70
Sag ponds, 260
Sahara Desert:
  causes of aridity, 200
  dunes, 213, 214, 215
  erg, 211
  sirocco, 209-210
St. Elias Mountains, Alaska, 248
St. François Mountains, 151
St. Michel d'Aiguilhe, Chapel of, 174
St. Paul Rocks, 337
St. Peter Rocks, 337
St. Pierre, Martinique, 172
Salinas, 204, 271
Salmon Mountains, Idaho, 148
Saltation, 209
Salt deposits, 203*
Salt domes, 117*-118, 193, 269
  Gulf of Mexico, 332
Salt lakes, 268-272
Salt licks, 117

Salt marshes, 313
San Andreas Fault, 121
  offset streams, 97, 118*
  sag ponds, 260
San Bernardino Mountains, Calif., 146
San Cristobal, Lake, 57, 262
Sand barrier, 97, 311*-313
Sand drifts, 210
Sandia Mountains, N. Mex., 120*, 377*
Sand ripples, 310
Sand shadow, 210-211
Sand sheets, 210
Sandstone, 38-39, 42*, 74*
  nature of, 41
Sand storms, 209
Sandy Hook, N. J., 313
San Francisco earthquake, 121
San Francisco Peaks, Ariz., 166
San Gabriel Mountains, Calif., 146
Sangre de Cristo Mountains, 141
San Joaquin River, Calif., 262
San Joaquin Valley fans, 193
San Juan Mountains, Colo., 166, 262
  dikes, 182-183
  mudflow from, 57-58
San Juan River, 97-98
San Luis Valley, Colo., 215*
San Rafael Swell, Utah, 116
Santa Ana Mountains, Calif., 86
Santa Ana River, 86
Saranac, Lake, N. Y., 128, 260
Saturated zone, 70
Sawatch Range, Colo., 141
Sawtooth Range, Idaho, 147*, 148
Scablands, 188-189
Scarps. See Cuestas; Fault- and Fault-line scarps; Hogbacks
Schist, 38-39, 43, 44, 45*
Schoharie Creek, N. Y., 89
Schooley Peneplain, 102
Schunemunk Mountain, N. Y., 113, 232*
Scoria, 41, 163, 167*
Scotts Bluff, Nebraska, 4*
Sea arches, 308
Sea bottoms, 325-348
  abyssal plains, 335-336
  faults, 328, 332
  map, 326
  Mid-Ocean Ridge, 336-338
  old ideas about, 325-326
  profiles compared to continents, 327*

INDEX · 370

Sea bottoms—*cont.*:
 relief of, 328
 rock of, 328
 submarine canyons, 333-335
 submarine plateaus, 331
 trenches, 332, 336
Sea cave, 303*
Sea level:
 as base level, 16
 post-Pleistocene rise, 316
 streams affected by changes, 97
 "swinging," 343
Seamounts, 338-344
Sea puss. *See* Rip currents
Seashores. *See* Shores
Sea stacks, 308, 319*
Sedimentary rocks, 38-39, 41-43
 see also Limestone; Sandstone; *etc.*
Sediments:
 geosynclines, 132-133
 sea bottoms, 324, 330-335, 346
 shores, 309-316, 320-321
 streams, 70-71, 76-78, 81
 see also Deserts; Glacial moraine
Seifs, 213*, 214
Seismic sounding, 327-328
Seismic waves, 18
Semisopochnoic Island, Alaska, 156
Seracs, 225, 227*
Serir, 208*
Sevier, Lake (Utah), 270
Seymour Narrows, B. C., 306
Shale, 38-39, 41*
Shari River, Africa, 203
Shasta, Mount, 167
Shawangunk Mountains, N. Y., 111*
Shenandoah River, 78*, 89*, 90-91
Sherburne, Lake (Mont.), 255*
Shield. *See* Continental shields
Shield volcanoes, 168*-169
Shingle, 309
Shore face, 310
Shores, 296-323
 erosion, 298-299, 313-316
 evolution, 316-317
 marine effects, 299
 terrestrial effects, 298-299
 see also Coasts
Shore trough, 310
Shoshone River, Wyo., 86
Sierra Madre, Mexico, 143
Sierra Nevada, Calif.
 diagram, 146*
 fault block, 121
 influence of jointing, 34

Sierra Nevada—*cont.*:
 lakes, 255
 parallel drainage, 93
Sierra Nevada, Spain, 143
Sifnos Island, 122*
Sigsbee Deep, 332
Sills, 159*, 179, 181
Silurian Period, 24
Silver Spring, Fla., 281, 295
Sink, valley (uvala), 279
Sink, volcanic, 162
Sinkhole pond, 264*, 265
 in quarry, 280*
 Mammoth Cave region, 291
Sinking creeks, 281
 see also Lost rivers
Siphons, 284
Sirocco, 209-210
Skjaldbreidur (volcano), 337*
Slate, 38-39
Slickensides, 120*, 121, 124
Slide Lake, Wyo., 262
Slides, 57*
Slip, in faulting, 119
Slopewash, 71
"Slumgullion" mudflow, 57, 262
Slumping, 56
Smith, William, 22
Snag Lake, Calif., 262
Snake River, 86, 140, 189
Snake River Plateau, 175-176
"Snowballs," gypsum, 293
Soda Lake, Nev., 272
Sogne Fiord, Norway, 233
Soil creep, 56
Solar system, origin of, 17
Solifluction, 56
Solution basins, 264*-265, 278, 279*, 282
Solution cavities, 50*, 277*
Solution pans, 278, 279*, 282
Solution process in limestone, 276-277
Solution valley, 282
Sonoran Desert, 145-146
Sorgue River, France, 281
Sounding, 326-327
South Downs, England, 116
Spatter cones, 161, 168
Spelunkers, 282, 284
Spine, volcanic, 163, 172
Spits, 212*, 313
Springs:
 artesian, 295
 desert, 203
 mountain, 70
 rises, 281
 volcanic regions, 165, 189
Spurn Head, England, 313, 316
Square Butte, Mont., 154

Stacks, sea, 319*
Stalactites, 283*-287
Stalagmites, 283* 287
Standing Rock Basin, Utah, 32*
Steamboat Rock, 9*
Steens Mountain, Nev., 268*
Steptoes, 188
Stillstand, 102
Stocks, 153, 159*, 170
Stone Mountain, Ga., 48, 155
Strabo, 8, 11
Strandflat, Norway, 316-317
Strath terraces, 98-99
Stratified rocks. *See* Sedimentary rocks
Streams, 65-103
 cold highlands, 73
 consequent, 82, 83*
 desert, 201*, 202-205
 drainage systems, 87-91
 genetic types of, 112*
 modes of origin, 83*
 obsequent, 83*
 piracy, 88-91
 resequent, 83*
 small, 61
 spacing of, 91
 subsequent, 83*
 temporary, 279
 work of, 59-61, 70
 see also Currents; Rivers
Striations, glacial, 231, 232*
Strike and dip, 110*
Strike valleys, 83
Structural benches, 27*, 75, 83, 98, 99
Submarine canyons, 333*-335
Submergence, shore of, 296*
Subsequent streams, 83*, 112*
 Appalachians, 84
 captures by, 90
Subsidence, streams affected by, 100
Subtropical deserts, 200
Sun, origin of, 17
Sundance Sea, 140
Sunset Crater, Ariz., 166
Sunspot activity, 249
Superior, Lake, 259
Superposed streams, 85*-86*
 Black Hills, 150
 captures of, 90
 Rocky Mountains, 140
Susquehanna River, 86
Swallow holes, 280, 281, 295
Swamps:
 basins made by plants and animals, 262-263
 defined, 253
 oxbow, 260
Swash, 303

INDEX · 371

Swash marks, 310
Swells, 116, 300-301, 306
Swinging, river, 78
Syenite, 38-39
Synclinal valleys, 83
Synclines, 14*, 107*, 108

Tablelands. See Mesas
Table Mountain, 190, 347*
Tablemounts. See Seamounts
Taconic schist and slate, 44
Tahoe, Lake (Calif.), 146, 259*
Talus, 53-54, 58*, 145*, 146
Tanana River, Alaska, 96
Tanganyika, Lake, 260
Tarim Desert, 144, 200
Tarn River, 279
Tasca, Lake (Minn.), 94
Taurus Mountains, Turkey, 143-144
Taylor, Mount (N. Mex.), 174
Teays River, 244
Temperature changes, 47
  see also Climate
Terminal moraine, 233, 239
Terraces:
  marine, 310-311, 323
  strath, 98-99
  travertine, 286-287
Tethys Sea, 144
Teton Range, Wyo., 131*
  a composite scarp, 127
  history of, 147
  horn peaks, 235
  Snake River superposed on, 86
Thingvellir graben, 337*
Thomson, Sir Wyville, 326
Thun, Lake (Switzerland), 262
Tidal bore, 306
"Tidal wave." See Tsunami
Tide marshes, 313, 314*
Tides in lakes, 323
Till, glacial, 195-196, 242
Till plains, 195-196, 197*
Timber-raft dam, 264
Timpanogos, Mount (Utah), 109*
Timpanogos National Monument, Utah, 285*
Toadstool rocks, 30*, 51*, 217*
Tom, Mount (Mass.), 167
Tombigbee River, Ala., 44*
Tombolo, 312*, 313
Tonga Trench, 336
Tossa de Mar, 2*
Tracy Arm fiord, 148*
Trade winds, 300-301
Transverse Ranges, Calif., 146
Transylvanian Alps, 143

Travertine, 284-289
Trellis drainage, 91, 92*
Trenches, submarine, 332, 336
Triassic Period, 24
Tributaries, 70, 76
  glaciers, 225
  see also Drainage systems
Trieste Deep, 336
Trim line, 233
Tristan da Cunha Island, 337
Trough. See Glacial troughs; Valleys
Trough lakes, 233, 255, 256*
Truncated spurs, 231, 236, 237*
Tsunami, 299
Tuamotu atolls, 343
Tucson Mountains, Ariz., 201*
Tuff, volcanic, 163, 170, 171*
Tulare, Lake (Calif.), 261
Tumuli, 162
Tundra lakes, 263
Turbidity currents, 329, 334-336
Turkestan Desert, 200

Ubehebe Crater, Calif., 28
Uinta Mountains, Utah, 141
Unaka Mountains, N. C., 103, 155
Unakas, 155
Unaweep Canyon, Colo., 100*
Uncomphagre National Forest, Colo., 100*
"Undertow," 306
Unloading, 46*, 47
Uplift:
  base level and, 16, 105-107
  mountain-building, 132-136
  postglacial, 246-247, 329
  streams affected by, 97-101
  see also Diastrophism
Upper Firth River, Alaska, 79*
Utah, Lake (Utah), 270
Uvalas, 279

Valais Alps, 142*
Valles Caldera, 171
Valley of 10,000 Smokes, 178
Valley of the Goblins, Utah, 217*
Valleys:
  anticlinal, 83
  blind, 281-282

Valleys—cont.:
  cut mostly by rivers, 67
  dry, 281-282
  glaciated, 220*, 222-226, 231, 237*
  hanging, 75*, 124, 231
  in fault scarp, 124*
  karst, 281-282
  longitudinal, 83, 110
  pocket, 281-282
  ramp, 121
  rift, 121 (see also Grabens)
  solution, 282
  strike, 83, 110
  structural, 279
  structural influences, 82-85
  synclinal, 83
Valley sinks, 279
Valley trains, 240
Venetz-Sitten, Ignaz, 222
Ventifacts, 206-207
Vesuvian Volcanic Observatory, 170
Vesuvius, Mount, 169*-170
  age of, 28
  eruption of A.D. 79, 11-12
  mudflow from, 57
Victoria Falls, Africa, 97, 121
Virgin Islands, 298*, 346*
Vitruvius, 11
Volcanic blocks, 162
Volcanic bombs, 162
Volcanic islands, 172-175
Volcanic necks, 173*-175*
  Le Puy, 13*
Volcanic tuff, 163, 170, 171*
  see also Lava dust
Volcanism, 157-183
  areas of, 106, 157-158
  causes, 19, 159
  coastal effects, 321-322*
  continents formed by, 17, 158
  early ideas about, 11-13
  islands created by, 338-344
  region of (diagram), 159*
  relics, 28, 167
Von Buch's views, 10
Volcanoes:
  age of, 28-29
  Alaskan, 1*, 156*, 167, 172, 339*
  anatomy of, 159-162
  Azores, 172-173
  Cascade Range, 166-167*
  cinder-cone type, 160*
  composite (strato-), 161
  distribution (map), 157
  ejecta, 162-165
  erosion on, 92, 165-167
  explosive types, 169-171

INDEX · 372

Volcanoes—cont.:
  extinct, 173-175
  glaciers on, 166
  Hawaiian "shields," 160
  Japanese, 173
  life history of, 161-162
  Mexican, 167
  mudflows on, 57
  North American, 166-167
  number of, 158
  origin of, 159
  shield type, 159-160,
    168*, 169, 337*
  sizes of, 157-158
  South American, 143
  studies of, 170
  see also Volcanic necks;
    Volcanism; etc.
Vosges Mountains, France,
  122
Vulcan, 11
Vulcano, 11

Walvis Ridge, 338
Warping, 108
  see also Folds
Wasatch Range, Utah, 146
  faceted scarp, 124
  fault block, 121
  Mount Timpanogos, 109*
  total displacement, 123
Washes, desert, 203
Washington, Mount
  (N. H.), 151
Watchung Hills, N. J., 167
Water:
  ground, 69-70
  juvenile, 178 .
  see also Erosion; Hydrologic cycle: Rivers;
    Streams; etc.
Waterfalls, 75*
  at nickpoints, 99
  faulting as cause, 97
  from hanging valleys, 231
  origin of 72, 76, 84
  over canyon walls, 204
  "petrified," 293
Water gaps, 64*, 86*
  coastal plains, 192
  nature of, 85-86

Water table, 70*
  karst terrain, 280-284
  nature of, 70
Waterton Lakes National
  Park, 141*
Watkins Glen, N. Y., 95*
Wave-cut bench, 303, 319*,
  321-322
  California, 321
  evolution of, 316
  Kerrera, 15*
  rarity of, 316-317
  relative to marine-cut
    terrace, 310-311
Wave-cut terraces, 323
Wave drift, 300
Wave refraction, 315*, 324*
Waves:
  breakers, 301-304
  erosion by, 302-304
  lake, 323
  mechanics of, 299*
  nature of, 299*-304
  swell, 300-301
  tsunamis, 299-300
  wind-made, 300, 301
Weald, The, 83, 92, 116-
  117*
Weathering:
  chemical, 50-52
  defined, 33
  desert, 205-206
  limestone caverns, 289
  mechanical, 47-49
  nature of, 31-32
  valley widening, 71-72
Wegener, Alfred, 345
Weisshorn, 144
Western Heights (Weald),
  116
West Indies, 125, 172
Westminster Abbey, 52
West Yellowstone earthquake, 55*, 97
Wetterhorn, Alps, 248*
Whalebacks, 214
White Mountains, N. H.:
  cirques, 236
  felsenmeer, 48*, 49
  Great Gulf, 152*
  influence of jointing, 34

White Mountains—cont.:
  intrusive rocks, 35, 174-
    175
  lakes, 255
  origin, 138-139, 151
  patterned ground, 56
  ring dikes, 174-175
White River, Ark., 80*
White Rock Mountain, Ark.,
  153*
White Sands National
  Monument, 212*, 216
Wight, Isle of, 116
Wind:
  effects in desert, 205,
    206-219
  lake basins made by,
    265-266
  loess distribution by, 245
  trade, 200-201
  see also Dunes; Waves
Wind Cave, S. D., 286*
Wind gaps, 90*
Windows, 207*, 281
Wind River Range, Wyo.,
  140

Xanthos of Sardis, 326

Yardangs, 208
Yazoo streams, 94
Yellowstone region:
  earthquake, 55*, 97
  igneous phenomena, 166,
    177*-178, 181*
  plateau, 188
  ponds, 262
Yellowstone River, Wyo., 84
Yorkshire (England) coast,
  316
Yosemite Valley, 75*, 234-
  235
Yukon Flats, 77, 95-96*,
  102

Zambesi River, Africa, 97
Zion National Park, Utah,
  42*, 154
Zuider Zee, 261
Zuni Uplift, N. Mex., 115,
  148